Nonlinear Physical Science

Nonlinear Physical Science

Nonlinear Physical Science focuses on recent advances of fundamental theories and principles, analytical and symbolic approaches, as well as computational techniques in nonlinear physical science and nonlinear mathematics with engineering applications.

Topics of interest in *Nonlinear Physical Science* include but are not limited to:

- New findings and discoveries in nonlinear physics and mathematics
- Nonlinearity, complexity and mathematical structures in nonlinear physics
- Nonlinear phenomena and observations in nature and engineering
- Computational methods and theories in complex systems
- Lie group analysis, new theories and principles in mathematical modeling
- Stability, bifurcation, chaos and fractals in physical science and engineering
- Nonlinear chemical and biological physics
- Discontinuity, synchronization and natural complexity in the physical sciences

Series editors

International Advisory Board

More information about this series at http://www.springer.com/series/8389

Rómulo Antão

Type-2 Fuzzy Logic

Uncertain Systems' Modeling and Control

With contributions by
Alexandre Mota
Rui Escadas Martins
José Tenreiro Machado

Higher Education Press

Springer

Author
Rómulo Antão
Department of Electronics,
Telecommunications and Informatics
University of Aveiro
Aveiro
Portugal

With Contributed by
Alexandre Mota
Department of Electronics,
Telecommunications and Informatics
University of Aveiro
Aveiro
Portugal

Rui Escadas Martins
Department of Electronics,
Telecommunications and Informatics
University of Aveiro
Aveiro
Portugal

José Tenreiro Machado
Institute of Engineering
Polytechnic of Porto
Porto
Portugal

Additional material to this book can be downloaded from http://extras.springer.com

ISSN 1867-8440 ISSN 1867-8459 (electronic)
Nonlinear Physical Science
ISBN 978-981-10-4632-2 ISBN 978-981-10-4633-9 (eBook)
DOI 10.1007/978-981-10-4633-9

Jointly published with Higher Education Press, Beijing
ISBN: 978-7-04-047809-9 Higher Education Press, Beijing

The print edition is not for sale in China Mainland. Customers from China Mainland please order the print book from: Higher Education Press.

Library of Congress Control Number: 2017941483

Printed on acid-free paper

This Springer imprint is published by Springer Nature
The registered company is Springer Nature Singapore Pte Ltd.
The registered company address is: 152 Beach Road, #21-01/04 Gateway East, Singapore 189721, Singapore

Preface

This book covers a new Fuzzy Logic domain—the Type-2 Fuzzy Logic, which has recently received an increasing attention from both academia and industry due to its advantages in handling information uncertainty in computational systems. More particularly, it covers a particular domain related to process modelling and control applications. It provides a new approach for those who seek to use in a single framework the advantages of model-based control algorithms, such as the possibility of developing high-performing closed-loop systems, and the superior model uncertainty manipulation capabilities of Type-2 Fuzzy Logic.

Its contents are presented in a bottom-up approach starting with the introduction of the fundamental concepts of Type-2 Fuzzy Sets, describing how process models can be easily developed according to its principles and, ultimately, integrate them in state-of-the art model-based control algorithms. Throughout the book, theory is complemented with practical applications and reader is invited to take his learning process one one-step forward by implementing his own applications using the book materials.

Aveiro, Portugal
December 2016

Rómulo Antão

Contents

Acronyms

ANFIS	Adaptive Network-based Fuzzy Inference System
ANN	Artificial Neural Network
API	Application Programming Interface
ARMAX	Auto-Regressive Moving Average with eXogenous input
ARX	Auto-Regressive with eXogenous input
CARIMA	Controlled Auto-Regressive Incremental Moving Average
CARMA	Controlled Auto-Regressive Moving Average
CE	Control Effort
CTS	Coupled Tanks System
DMA	Direct Memory Access
DMC	Dynamic Matrix Control
EODS	Enhanced Opposite Direction Searching Algorithm
FCM	Fuzzy C-Means
FL	Fuzzy Logic
FLS	Fuzzy Logic System
FM	Fuzzy Model
FOU	Footprint of Uncertainty
FPU	Floating Point Unit
FS	Fuzzy Set
GPC	Generalized Predictive Control
HIL	Hardware-In-the-Loop
IT1FS	Interval Type-1 Fuzzy Set
IT2FLS	Interval Type-2 Fuzzy Logic System
IT2FS	Interval Type-2 Fuzzy Set
KM	Karnik-Mendel
LMF	Lower Membership Function
MAC	Model Algorithmic Control
MCU	Micro-Controller Unit
MF	Membership Function
MIMO	Multiple-Input Multiple-Output

MPC	Model Predictive Control
MSE	Mean Squared Error
NARX	Nonlinear Auto-Regressive with eXogenous input
NFS	Neuro Fuzzy System
NSE	Normalized Squared Error
PIL	Processor-In-the-Loop
RBFN	Radial Basis Function Network
RLS	Recursive Least Squares
RLSDF	Recursive Least Squares with Directional Forgetting
SISO	Single-Input Single-Output
SNR	Signal-to-Noise Ratio
TR	Type-Reduction
TS	Takagi-Sugeno
UMF	Upper Membership Function

Chapter 1
Introduction

The development of flexible control algorithms to manipulate a process close to its best performance without human supervision has been one the most sought goals of control and system modeling theory. The work of Zadeh during the early 60's in the theory of Fuzzy Sets(FSs) and Fuzzy Logic (FL) introduced a fundamental degree of fuzziness in computational systems. That strategy overcame the difficulties of obtaining accurate descriptions of models and control systems due to the inherent variability and noisy operation conditions of real world processes. His achievements significantly contributed to the state-of-the-art of current technology in a broad range of applications and even today, on the 50th anniversary of FL's seminal work [1], continue to bring about new approaches to optimize the way information uncertainty is accounted for in computational systems. The Type-2 FL is one of its most recent extensions.

Despite invariably linked with information fuzziness, the original FL theory does not consider the inherent uncertainty of assigning a single membership function to each FS defined over a numerical domain—each membership function chosen is itself crisp since it is totally defined without considering any variability on its parameters (such as its center, width, endpoints or shape). The Type-2 FS overcome this limitation by introducing additional degrees of freedom in the membership function concept so higher levels of uncertainty over its representation are accounted for. Ultimately, a Type-2 FS embeds itself a large number of Type-1 FSs under the same label yielding a blurred Fuzzy Set representation.

Inspired by the simplicity of developing rule based systems, Fuzzy Logic Systems (FLSs) (based on the Mamdani or Takagi-Sugeno (TS) structures) were naturally improved by introducing the Type-2 FL formalisms to accommodate higher levels of uncertainties in the system's parameters. This transition is fairly natural since the basic principles of FL are independent of the nature of the membership functions, requiring only little changes in the typical FLS structure. As a elemental part of a FLS, Type-2 FSs provides a better coverage of the crisp domain of interest and ultimately contribute to the reduction of the number of rules required to approximate complex input-output data relationships [2]. While the additional degrees of freedom of Type-2 FSs shown greater potential to override conventional information representation

R. Antão, *Type-2 Fuzzy Logic*, Nonlinear Physical Science,
DOI 10.1007/978-981-10-4633-9_1

methodologies [3, 4], especially in complex scenarios described by non-linear data dependencies, generally their use requires a larger computational effort due to complexity of the required calculations. For that reason, a great amount of research in Type-2 FL domains has been put towards developing more efficient representations such as Interval Type-2 FSs [5], in order to get around this bottleneck and further extend its applicability to real world scenarios. Such approach is already yielding promising results, leading to successful applications in traditional FL domains such as modeling, control or classification systems.

As far as control systems development is concerned, the state-of-the-art of Type-2 FLSs does not seem to be taking full advantage of the most important achievements of model based control algorithms. Literature mainly highlights approaches based on PID structures [6–8]—whose discrete-time implementations still have deep roots in traditional continuous-time concepts such as step response analysis. One can also find model-based approaches using Direct Inverse models, obtained by means of analytical methods that directly invert Type-2 TS Fuzzy Models(FMs) [9, 10] or by modeling the inverse dynamics of the system [11]. While inverse model controllers are intuitively simple and eventual steady-state errors can be compensated by integrating the inverse model in an Internal Model Control structure [10], such approaches may not work satisfactory when a system's inverse model is not well-damped. To some extent, Type-2 Fuzzy Control state-of-the-art is not considering the improvements brought by model-based control design techniques such as Pole-Placement [12] or Model Predictive Control (MPC) [13]. The latter approach became, in fact, one of the most popular methods in both industrial and academic communities, efficiently handling a wide range of control problems with large number of design variables such as systems with multiple control inputs and control signal constraints. One of its simplest, yet robust implementations is the Generalized Predictive Control (GPC) algorithm [14].

The process's model is a cornerstone of every MPC implementation and its accuracy ultimately defines the quality of the control system in terms of tracking capabilities and robustness to external disturbances. While most of time linear approximations are enough, in some applications it is of uttermost importance to develop models that take in account the possible non-linearities of the process. Traditionally, MPC implementations are based on linear models but, in order to extend its theory to non-linear processes, the combination of FMs with MPC became increasingly discussed in recent years and has been the object of important studies regarding its stability and applicability [15–17]. More particularly, TS FMs shown advantageous for such purpose by two main reasons:

- Capability of modeling complex non-linear processes using input-output data along with *a priori* knowledge of the system provided by the user. By combining the efficiency of fuzzy reasoning in handling uncertain information and the neural networks learning ability in model's development, TS Fuzzy systems retain an important level of interpretability and adaptability in their structure.

- Its structure follows a two-layered computing scheme which partitions a non-linear system as a contributions of several locally linear models. Such topology avoids the use of extensive non-linear optimization algorithms during model training and allows the design of controllers according to linear control theory.

Type-2 FL is formal extension of its original Type-1 counterpart and shares many of its applications but, up to the date, examples of the use of Type-2 TS FMs as support for MPC algorithms are scarse [18]. Hence the main goal of this book is to present a systematic methodology to merge both domains and assess the performance improvements achieved over classical implementations of MPC, developing a closed loop control algorithm based on MPC theory and a Type-2 FMs which can be implemented in general purpose embedded systems.

In this line of thought, this book aims to expand the reader's knowledge of Type-2 FL in the following four topics:

- Develop simpler methods for training a Type-2 Takagi-Sugeno FM. While currently every parameters of such structure are trained as part of a single error minimization problem, it is computationally more efficient and tractable to the user to consider two separate problems: the training of a supporting Type-1 Takagi-Sugeno Fuzzy System and the introduction of a Footprint-of-Uncertainty (FOU) over the respective parameters. The width of the latter can be adjusted so the approximation capabilities of the model are improved.
- Apply the principles of Type-2 FL to reduce the influence of modeling uncertainties on locally linear prediction models. The development of a multi-step predictor for a non-linear system typically establishes a trade-off between accuracy and computational complexity. While a good compromise can be usually achieved using locally linear approximations from TS FMs, in changing operation regimes the predictor's validity may be significantly reduced. Though, its performance can be improved by obtaining a linearized model from a Type-2 Takagi-Sugeno FM and so the necessary procedures will be presented.
- Create model-based control algorithms according to the GPC principles using Type-2 FS. By synthesizing a control law based on linearized Type-2 TS FMs, a superior closed loop tracking performance and robustness to unmodeled operation conditions can be achieved comparatively to traditional GPC implementations.
- Implement a closed loop controller based on GPC theory and a Type-2 TS FMs in embedded platforms. The higher computational requirements of Type-2 FL based systems impose significant constraints over their use in real-time applications. Therefore, the algorithm's turnaround time will be evaluated when executed in a ARM® Cortex®–M4© microcontroller in order to assess its possible applications and limitations.

1.1 Book Outline

The contents of this book spans over seven chapters, organized as follows:

- Chapter 2 introduces the fundamental concepts of FLS and their extension to a particular branch of the Type-2 FL theory based on Interval Type-2 Fuzzy Sets.
- Chapter 3 focuses on a particular implementation of the fuzzy inference mechanisms, the TS Fuzzy Systems, presenting the improvements of its traditional formulation (based on Type-1 Fuzzy Sets) according to the most recent developments on Type-2 Fuzzy Sets' theory. The procedures for system identification based on the proposed structure are outlined.
- Chapter 4 presents an approach for the development of a n-step ahead prediction model based on the linearization of a Type-2 Takagi-Sugeno FM. The capabilities of such methodology are evaluated using two non-linear processes as benchmark systems.
- Chapter 5 proposes the use of Interval Type-2 FMs for the development of a MPC according to the GPC theory. Based on the two benchmark scenarios presented in the previous chapter, the performance of the respective closed loop control systems will be evaluated under several unmeasured external disturbances.
- Chapter 6 presents a Processor-in-the-Loop framework based on a ARM® Cortex®-M4 development board and the MATLAB® Simulink®. The proposed system is implemented not only to evaluate the feasibility of implementing Type-2 FLS in computationally constrained platforms, but also to improve the development and testing stages of complex embedded systems, providing an easier transition between the simulation and real world environments.
- Finally Chap. 7 presents the concluding remarks regarding the proposed methodologies, discussing some possible lines of work for future development.

References

1. Zadeh, L.: Fuzzy sets. Inf. Control **8**, 338–353 (1965)
2. Wu, D., Tan, W.: Type-2 FLC modeling capability analysis. In: Proceeding of the 2005 IEEE International Conference on Fuzzy Systems, pp. 242–247 (2005)
3. Wagner, C., Miller, S., Garibaldi, J., Anderson, D.: From interval-valued data to general Type-2 fuzzy sets. IEEE Trans. Fuzzy Syst. **23**(2), 248–269 (2015)
4. Wagner, C., Hagras, H.: Towards general Type-2 fuzzy logic systems based on zSlices, IEEE Trans. Fuzzy Syst. **18**(4) (2010)
5. Liang, Q., Mendel, J.: Interval Type-2 fuzzy logic systems: theory and design. IEEE Trans. Fuzzy Syst. **8**(5), 535–550 (2000)
6. Wu, D., Tan, W.: A simplified Type-2 fuzzy logic controller for real-time control. ISA Trans. **45**(4), 503–516 (2006)
7. Hagras, H.: Type-2 FLC: A new generation of fuzzy controllers. IEEE Comput. Intell. Mag. **2**(1), 30–43 (2007)
8. Kumbasar, T; Hagras, H.: A gradient descent based online tuning mechanism for PI type single input interval Type-2 fuzzy logic controllers. In: Proceedings of IEEE International Conference on Fuzzy Systems (2015)

9. Kumbasar, T.; Eksin, I.; Guzelkaya, M.; Yesil, E.: Type-2 fuzzy model inverse controller design based on BB-BC optimization method. In: The 18th World Congress of the International Federation of Automatic Control (IFAC) (2011)
10. Kumbasar, T., Eksin, I., Guzelkaya, M., Yesil, E.: Interval Type-2 fuzzy inverse controller design in non-linear IMC structure. Eng. Appl. Artif. Intell. **24**, 996–1005 (2011)
11. Zhao, L.: Direct-inverse modeling control based on interval Type-2 fuzzy neural network. In: Proceedings of the 29th Chinese Control Conference, pp. 2630–2635 (2010)
12. Ljung, L.: System Identification: Theory for the User. Prentice Hall (1987)
13. Camacho, E., Bordons, C.: Model Predictive Control, Advanced Textbooks in Control and Signal Processing, 2nd edn. Springer, London (2007)
14. Clarke, D., Mohtadi, C., Tuffs, P.: Generalized predictive control, Parts 1 and 2. Automatica **23**, 137–160 (1987)
15. Mollov, S., Babuŝka, R., Abonyi, J., Verbruggen, H.: Effective optimization for fuzzy model predictive control. IEEE Trans. Fuzzy Syst. **12**(5), 661–675 (2004)
16. Mendes, J.: Computational Intelligence Methodologies for Control of Industrial Processes, Ph.D. Thesis, Department of Electrical and Computer Engineering, University of Coimbra (2014)
17. Huang, Y., Lou, H., Gong, J., Edgar, T.: Fuzzy model predictive control. IEEE Trans. Fuzzy Syst. **8**(6), 665–678 (2000)
18. Antão, R., Mota, A., Martins, R.: Model-based control using interval Type-2 fuzzy logic systems. Soft Comput. Springer (2016)

Chapter 2
Fuzzy Logic Systems

2.1 Introduction

The human mind has always shown a remarkable capability of coordinating a wide variety of physical and mental tasks without using any explicit measurements and computations. Considerable efforts were made since the early 1950s towards the development of a scientific theory of intelligence and the development of an artificial model of the brain capable to mimic our perception, cognition and behavioral systems [1, 2]. Despite the accomplishments of system's theory and artificial intelligence, that are increasingly present in our daily activities, in practice computational systems still present several limitations that keep them behind human capabilities. The high dimensionality of information structures stored in computational systems resulting from the use of crisp measurements is one of the major burdens that the development of an intelligence framework must overcome. Uncertainties and imprecisions on information can at first instance be seen as a disadvantage for a decision process but they are important information compression mechanisms that let people make choices in a quick way. Without such tools, taking a decision would be a never ending process, requiring every infinitesimal part of information and its respective combinations to be considered. Therefore, the development of intelligent systems has to focus on the human capability of manipulating imprecise, uncertain and sometimes incomplete information.

Zadeh was challenged by this problem in 1965 and, in his seminal paper [3], he lays the foundation-stone of a methodology known as FL, where the objects of computation are words and propositions drawn from natural language. While Boolean logic results are restricted to values "0 and 1", FL defines for the first time a computational framework to efficiently manipulate intermediate results between the values of absolute true and absolute false. The fuzzy information representation is based on Fuzzy Sets, which are no more than a simple way to translate a crisp measurement into a degree of belonging in a linguistic label. This means that fuzzy

R. Antão, *Type-2 Fuzzy Logic*, Nonlinear Physical Science,
DOI 10.1007/978-981-10-4633-9_2

sets can handle some concepts that we commonly deal with in daily life, like "very cold", "cold", "hot", "very hot", without having to know the specific temperature ranges each concept refers to. Therefore, Fuzzy Logic is more like human thinking because of its reliance on degrees of truth and the use of linguistic variables.

Initially, FL theory was not well-received by the peer community in engineering domains due to its unusual vagueness. Nonetheless, since 1970, it has been widely applied in control applications, establishing successive milestones. Its principles were used to control a laboratory-built steam engine by Mamdani at the University of London in 1974 [4] and the first industrial application was a cement kiln controller built in Denmark in 1979 [5]. Despite born in the USA and theoretically validated in Europe, it was in Japan that FL gained broad notoriety when several Japanese companies pioneered successful practical applications with high impact in the society. One of the most renowned projects was presented in 1987, when Hitachi turned over control of a subway in Sendai, Japan, to a fuzzy system (Fig. 2.1). Fuzzy control techniques were used in all the critical operations of the train's control system, such as accelerating, breaking, and stopping operations [6] but also in traffic planning and predicting customer's usage of subway facilities. In 1987, Yamakawa successfully developed a fuzzy controller applied to a inverted pendulum experiment—a classic control problem [7]. A few years later, NASA took fuzzy logic beyond our planet aboard the Endeavour space shuttle, transporting a Commercial Refrigerator Incubator Module as an experimental payload, which successfully allowed the control of a test chamber's air temperature according to a pre-programmed profile [8]. Since then, several companies have been using fuzzy logic to control hundreds of household appliances, implement decision making systems and improve the performance of many other electronic devices present in our daily life such as air conditioners, video cameras, televisions, washing machines, bus time tables, medical diagnoses or anti-lock braking systems.

Fig. 2.1 Sendai subway 1000 Series—The first subway coach using a fuzzy control system

Fuzzy Logic Systems are typically developed around two main types of inference frameworks: the Mamdani [9] and the Takagi-Sugeno [10] methods. Despite the differences between the two methods, when non-linear Fuzzy Sets are used to model linguistic labels, FLSs become non-linear structures with universal approximation capabilities [11], a property of major importance when they are used as support for modeling and control techniques. Since both inference mechanisms share several theoretical principles, this chapter will firstly introduce FL's fundamentals based on the Mamdani inference.

2.2 Type-1 Fuzzy Sets

With the development of FL, the Type-1 FSs were defined for the first time. Type-1 FSs are a computational formalism that mimics our tendency to group crisp measurements displayed under a numeric scale using the same linguistic term when a more specific distinction is not required for a good understanding. For example, we describe temperatures using a linguistic terms that go from "Very Cold" to "Very Hot", speed using terms from "Very Slow" to "Very Fast" or the visible colors from "Violet" to "Red". With these descriptions, we are capable of abstain ourselves from the crispness of numeric scales expressed in °C, km/h or nm. Each Type-1 Fs is syntactically represented by a label F_i characterized by a Membership Function (MF), a two-dimensional function that defines the degree of association of a numeric value under the respective linguistic label using a crisp number in the range [0–1]. Different shapes of MFs can be considered, namely triangular, trapezoidal or gaussian shaped functions. Figure 2.2 depicts an example of a generic input domain partitioned using gaussian shaped MFs.

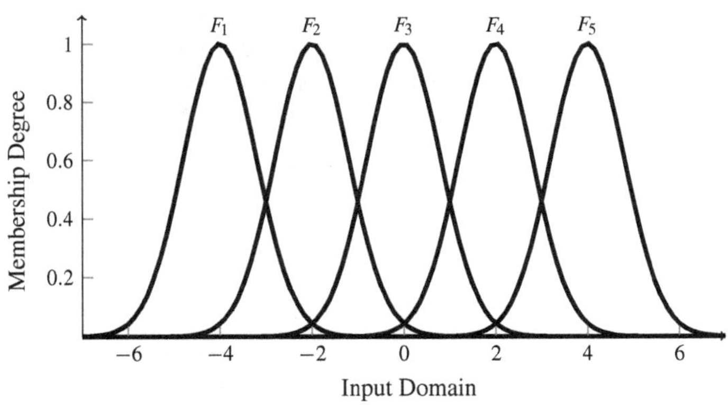

Fig. 2.2 Generic input domain partition using Type-1 Fuzzy Sets

2.3 Type-1 Fuzzy Logic Systems

One of the most appreciated methods of manipulating linguistic information is to use FLSs based on *If-Then* rules, a method that can be easily used to develop models and control algorithms in a way closer to human perception and thinking. There exist also alternatives to the use of rule based systems [12] involving arithmetic approaches using the Extension Principle [13]. To enunciate the required logic statements, this principle redefines common algebraic operations such as addition, multiplication, among many others to the domain of Fuzzy Sets. Sometimes it is hard to define *If-Then* rules using compact algebraic operations. Nevertheless, this approach is particularly useful in situations where the problem has a high dimensionality, i.e. the number of existing rules used to describe the system is so high that it may result in a computationally inefficient process.

A FLS based on Type-1 FSs consists of four main elements, as depicted in Fig. 2.3 and briefly described as follows.

- The Fuzzifier, which is an interface which maps a crisp number into a fuzzy domain defined by a Fuzzy Set. The most widely method is the *singleton fuzzifier*, in which measurements are considered perfect and therefore modeled as crisp values, i.e., as singletons.
- The Rule-Base, which is the heart of a FLS and is composed by information given by experts or extracted from numerical data, is often organized as several *If-Then* statements, where the *If* part of a rule is its antecedent, and the *Then* part of the rule is its consequent.
- The Inference Engine, which is the mechanism that implements the algebras required to manipulate Fuzzy Sets. In the same way humans use many inferential procedures, there exist several methods to do so based on FL. The Mamdani and the Takagi-Sugeno inference mechanisms are the two most popular ones [14].
- The Output Processor, which is the final stage of the FLS and implements the defuzzification procedures to aggregate the output fuzzy set into a single crisp value adequate to the FLS application scenario (usually a process's output prediction in modeling applications or the actuation value of a control system).

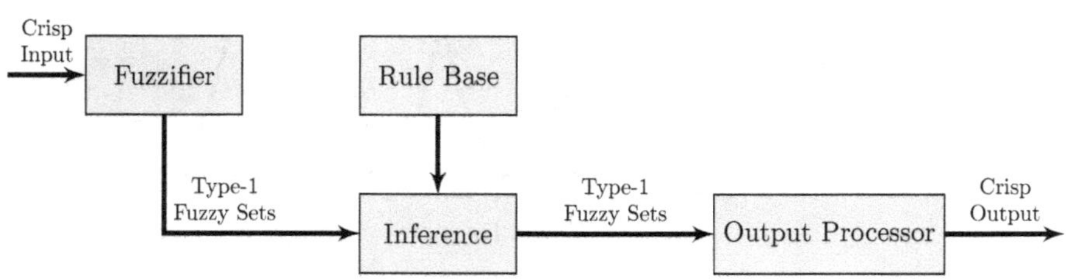

Fig. 2.3 Type-1 fuzzy Logic System structure

2.3.1 Fuzzifier

As pointed out previously, the singleton fuzzifier is the most used method to implement the Fuzzifier stage of a FLS due to its conceptual simplicity and easy data manipulation in the subsequent stages of the FLS. Other functions could be used to perform this operation, as gaussian or triangular functions, but then calculating the firing levels of each antecedent membership function would become a far more complex process [13]. For this reason, singletons are considered in this work, and defined as:

$$\mu_{A_x}(x) = \begin{cases} 1, & x = x' \\ 0, & \text{otherwise} \end{cases}, \tag{2.1}$$

where x' is the input value. This concept is depicted in Fig. 2.4.

2.3.2 Rule-Base

Rules play a central role in the structure of a FLS and are a simple way to gather the knowledge that define the behavior of a fuzzy system in a specific application. These rules are developed using different types of FSs, that are associated with the linguistic terms that appear in the rule's antecedent and consequent parts and are interconnected by operators that establish the relationship dependencies between fuzzy terms.

The most used rule structure is presented in Eq. (2.2)

$$R^i : If\ x_1\ is\ F_1^i\ and\ \cdots\ and\ x_j\ is\ F_j^i,\ Then\ y^i\ is\ G^i\ , \tag{2.2}$$

where R^i represents the ith fuzzy rule, F_j^l and G^i are linguistic terms characterized by Type-1 Fuzzy Sets, $i = \{1, \cdots, M\}$ where M is the number of fuzzy rules,

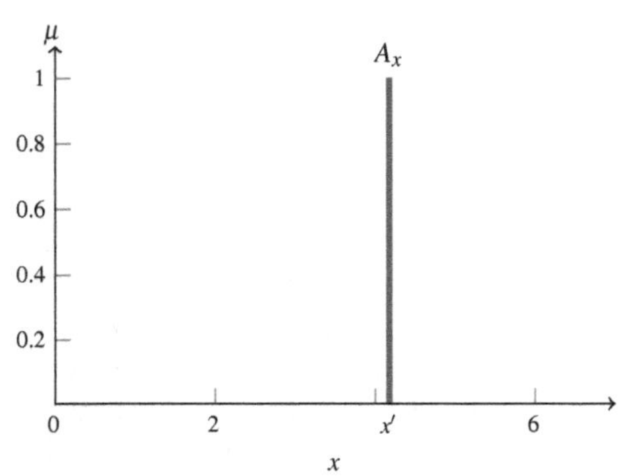

Fig. 2.4 Depiction of a singleton input

$j = \{1, \cdots, N\}$ where N is the number of antecedents, x_j are the fuzzy system inputs and y^i is the rule output.

The linguistic terms F and G can assume several different shapes such as triangular, trapezoidal or gaussian. The latter form was employed in this work, being defined as presented in Eq. (2.3).

$$g(c, \sigma, x) = \exp\left[-\frac{1}{2}\left(\frac{x-c}{\sigma}\right)^2\right] \tag{2.3}$$

The rule structure (2.2) is just one example of many prospective ways to embed knowledge in a FLS. Similarly to our natural language, one can employ other combinations of FSs using for example *or* relationships, consider the negation of fuzzy sets, or even using *non-obvious* connectives like *unless* or comparative terms [13]. Though, in most scenarios, such logical statements can be represented using the more regular structure of Eq. (2.2).

2.3.3 *Inference Engine*

The manipulation of fuzzy sets can be performed according to several different algebraic operations depending on the possible combinations of operators used for implementing the rule's connective terms, implication methods and rule aggregation. Let us consider two Type-1 FS F_1 and F_2 characterized by the MFs $\mu_{F_1}(x)$ and $\mu_{F_2}(x)$

$$F_1 = \int_{x \in \mathbb{R}} \mu_{F_1}(x)dx\ , \qquad F_2 = \int_{x \in \mathbb{R}} \mu_{F_2}(x)dx\ . \tag{2.4}$$

The basic logic operations (union, (*s-norm*), intersection (*t-norm*) and complement (*c-norm*)) that provide the support for FS's manipulation can be defined as follows:

$$\begin{aligned} \mu_{F_1 \cup F_2}(x) &= max\left[\mu_{F_1}(x), \mu_{F_2}(x)\right],\ for\ x \in \mathbb{R} \\ \mu_{F_1 \cap F_2}(x) &= min\left[\mu_{F_1}(x), \mu_{F_2}(x)\right]\ , \\ \mu_{\overline{F}}(x) &= 1 - \mu_F(x)\ . \end{aligned} \tag{2.5}$$

The intersection operator can also be implemented based on the algebraic product and is defined as:

$$\mu_{F_1 \cap F_2}(x) = \mu_{F_1}(x) * \mu_{F_2}(x),\ x \in \mathbb{R}\ . \tag{2.6}$$

Since the usual way of constructing a rule is to use the *and* connectives, the *t-norm* operators are the most used ones. Regardless the implementation followed, since the membership grades $\mu_{F_1}(x)$ and $\mu_{F_2}(x)$ are crisp numbers, any of the operations presented in Eqs. (2.5) and (2.6) yields a crisp number. More particularly, when these

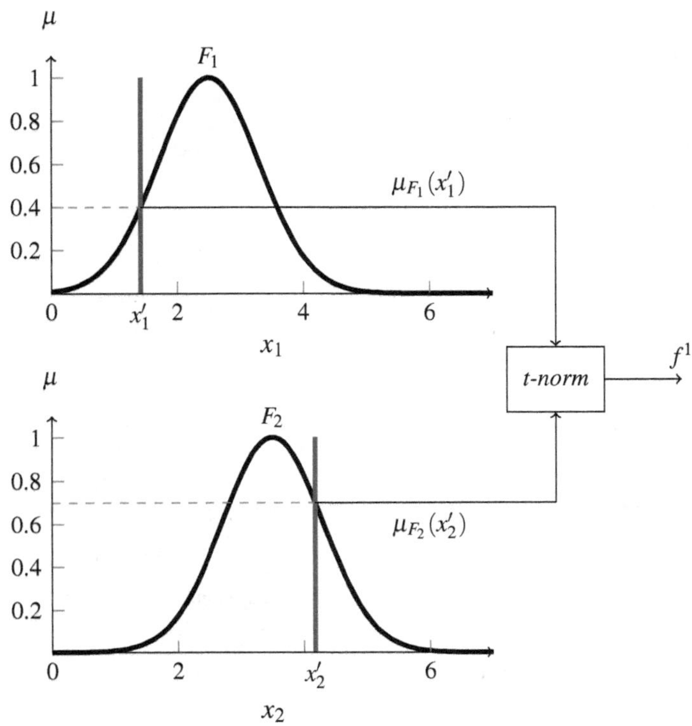

Fig. 2.5 Operation between singleton input and the antecedents of a Type-1 FLS using a *t-norm* operator (minimum or product)

operators are used to aggregate several fired FSs from the antecedent part of the rule, the obtained result is typically referred to as the ith rule firing level f^i. Figure 2.5 depicts the use of the *t-norm* in a two antecedent operation.

When both rule's antecedents and consequent are expressed using Type-1 FS, the implication of the rule's firing level over the consequent FS is typically obtained using one of the Mamdani implication methods, namely the Mamdani minimum or the Mamdani product. These methods are based on the *t-norm* operators previously described and are applied between the rule's firing level f^i and its consequent FS, $G^i(x)$

$$\begin{aligned} f^i &\rightarrow G^i(y) = min\{f^i, \mu_{G^i}(y)\},\ \ for\ y \in \mathbb{R} \\ f^i &\rightarrow G^i(y) = f^i \mu_{G^i}(y)\ . \end{aligned} \tag{2.7}$$

Depending whether the minimum or the product *t-norm* is used, one obtains a clipped or a scaled version of the consequent MF, as depicted in Fig. 2.6.

The inference process ends up by obtaining a fuzzy set determined by the aggregation of the output of all the fired FLSs rules. One of the most used methods to do so is to use a *t-conorm*—the fuzzy union, which is no more than a method for finding the maximum value of the overlapped FSs. This is depicted in Fig. 2.7.

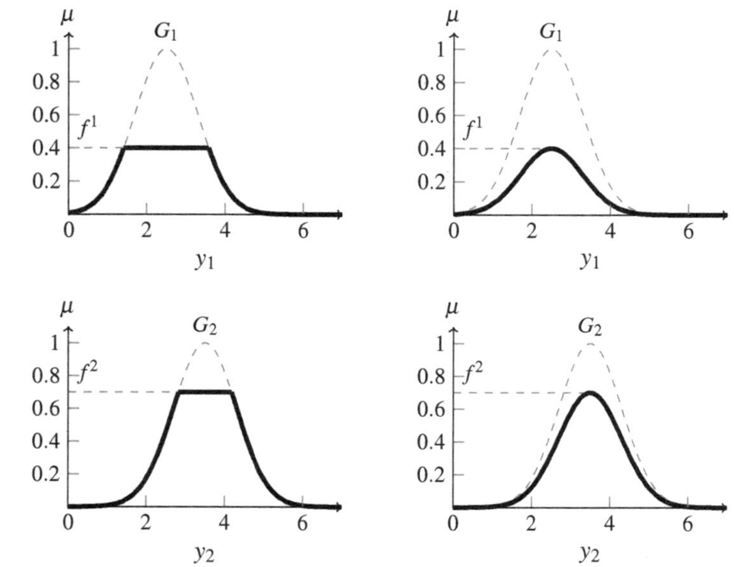

(a) Fired output sets using the minimum *t-norm*.

(b) Fired output sets using the product *t-norm*.

Fig. 2.6 Mamdani inference operations using Type-1 FSs

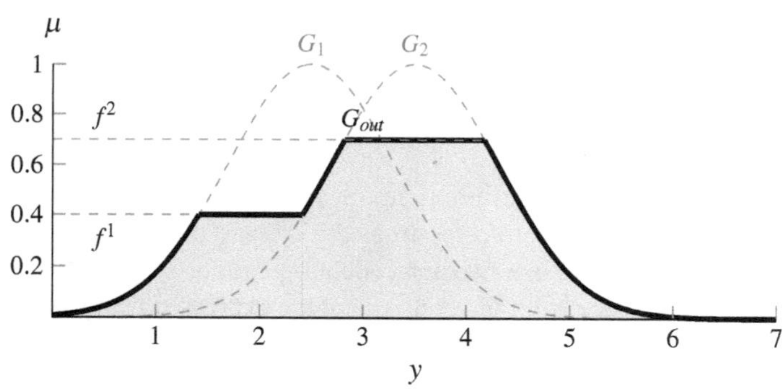

Fig. 2.7 Type-1 Fuzzy Set fired consequents' aggregation procedure, after using the mamdani minimum implication

Still, this final step is not consensual among authors, existing some that give preference to aggregate the output of every rule before defuzzification while others perform the aggregation as part of the defuzzification procedure. Given this fact there exist several different methods to implement FLS defuzzification stage, as will be following discussed.

2.3.4 Output Processor

The Output Processor is employed to obtain a crisp output from the FS resulting from the inference procedure, a process known as defuzzification. The literature is rich in defuzzifiers methods but, considering applications in modeling and control domains, their computational efficiency is often the main exclusion criterion. From the available methods, the Centroid and the Center-of-Sets defuzzifiers are the most frequently used ones [13], and represent good examples of the two different aggregation/defuzzification approaches.

The Centroid defuzzifier obtains a crisp value by finding the centroid of a Type-1 FS resulting from the union of the consequent fuzzy sets. By sampling the resulting G_{out} FS into K points, as depicted in Fig. 2.8, its centroid is given by:

$$y_C = \frac{\sum_{i=1}^{K} y_i \mu_{G_{out}}(y_i)}{\sum_{i=1}^{K} \mu_{G_{out}}(y_i)} . \tag{2.8}$$

Alternatively, the Center-of-Sets defuzzifier does not rely on prior aggregation of every fired consequent FS. Instead, it replaces every rule's consequent Type-1 FS by a singleton placed at its centroid which amplitude is given by the respective rule's firing level. The defuzzifier output value is then obtained by calculating the centroid of the Type-1 FS comprised of these singletons, and is given by:

$$y_{COS} = \frac{\sum_{i=1}^{M} c^i f^i}{\sum_{i=1}^{M} f^i} , \tag{2.9}$$

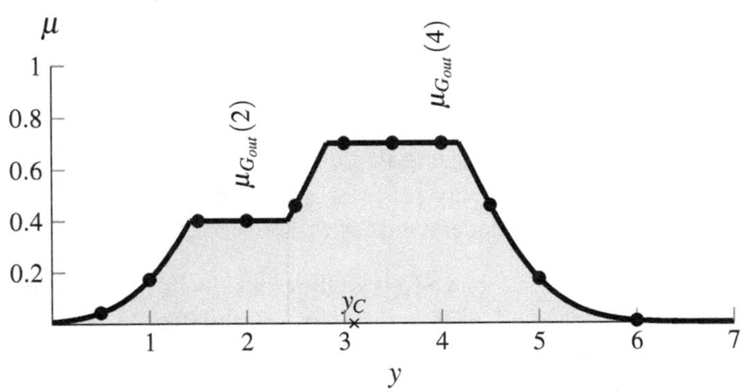

Fig. 2.8 Defuzzification procedure based on the centroid method applied to the G_{out} fuzzy set

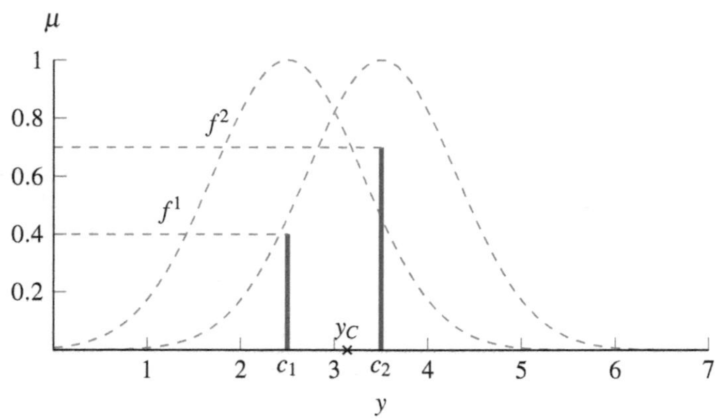

Fig. 2.9 Defuzzification procedure based on the Center-of-Sets method, using the centroid of each fired consequent fuzzy sets separately

where c^i is the centroid of the ith consequent FS, M is the number of rules of the FLS and f_i is each rule's firing level. This procedure is depicted in Fig. 2.9.

The latter approach is usually preferred when implemented in computationally constrained systems since the centroid of each consequent's FS can be calculated *a priori* and stored in the system, before the fuzzy system is deployed. After performing this step, the FLS's output is obtained as a weighted average of the pre-calculated centroids.

2.3.5 Considerations About Type-1 Fuzzy Logic Systems

The importance of the additive structure that a Fuzzy Logic System presents goes far beyond its rules' intelligibility. By using several rules, one is in fact defining several fuzzy patches and average the ones that overlap, ultimately performing a multi-variable function approximation. Such procedure's accuracy improves as the fuzzy patches grow in number and shrink in size.

The Type-1 FSs have been found to provide good results in uncertainty modeling, but there are several opinions referring that using them as a model for a linguistic label is an incorrect scientific theory [15]. As is pointed out in Jerry Mendel's line of reasoning [15], it is easy to understand the reasons for this refutation:

- A Type-1 FS representation for a word is well-defined by its Membership Function that is totally certain once all of its parameters are specified.
- Words mean different things to different people and so are uncertain.
- Therefore, it is a contradiction to say that something that is certain can model something that is uncertain.

The same way the variance provides a measure of dispersion about the mean, the uncertainty of a linguistic term also needs to be captured, and is not possible to represent such when a single static MF is used. In Fuzzy Set's theory, this second order uncertainty can be modeled by Type-2 Fuzzy Sets.

2.4 Type-2 Fuzzy Sets

The concept of Type-2 FSs was developed by Zadeh in 1975 as an extension of Type-1 FSs [16], but it only gained broader audience much more recently with the several developments proposed by Mendel and Karnik [17]. Type-1 FSs introduced an important degree of fuzziness to create a linguistic partition of a crisp domain. Nonetheless, the MFs used to do so are themselves crisp since they are totally defined without considering any uncertainty on their parameters. Type-2 FS overcome this limitation by defining a secondary degree of fuzziness. In this case, the membership value for each input of a FS is itself defined as a FS in the [0, 1] domain [13]. To better understand this new dimensionality, suppose the process of defining a concept as a Type-1 FS by polling a group of experts. After gathering all the responses, certainly it will be noticed that the endpoints of the membership function will vary from person to person. The union of all embedded Type-1 FSs eventually will end up in a blurred area, known as FOU, that is bounded by two MFs, namely the Upper Membership Function (UMF) and the Lower Membership Function (LMF). Furthermore, each membership function given by a person can be assigned with a variable weight according to the amount of confidence associated to its opinion, defining this way the secondary degree of fuzziness. For this reason, a Type-2 FS representation embeds additional degrees of freedom which can better handle uncertainties caused by noisy data and changing environments as is required for example when developing a process's model. Figure 2.10 gives a better overview of the new concepts introduced by Type-2 FSs, which can be generically represented by:

$$\tilde{F} = \int_{u \in X} \mu(u) du = \int_{u \in X} \int_{x \in J_x} g(x) dx du \qquad J_x \subseteq \Re, X \subseteq [0, 1], \tag{2.10}$$

where the $g(x)$ is one of the possible primary MFs, x is the FS input value and u is the primary membership degree.

Until the late nineties, research work on Type-2 FSs was of highly mathematical and theoretical nature, having few publications dedicated to it [18]. The main investigation line was focused on the development of logical operators, with important works from Mizumoto and Tanaka [19], Dubois and Prade [20] and more recently Karnik and Mendel [17]. Another important topic that has received little attention in the literature is the process of acquisition of the Type-2 FSs' membership functions. From the few works published about this topic, Turksen [21] proposes that a

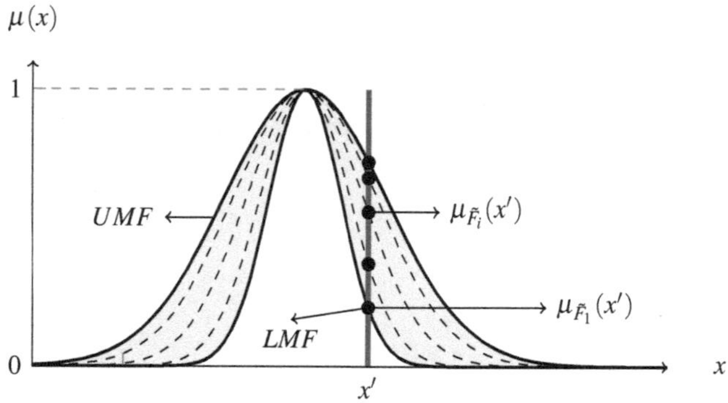

(a) FOU of a Type-2 Fuzzy Set, evincing several embedded Fuzzy Sets.

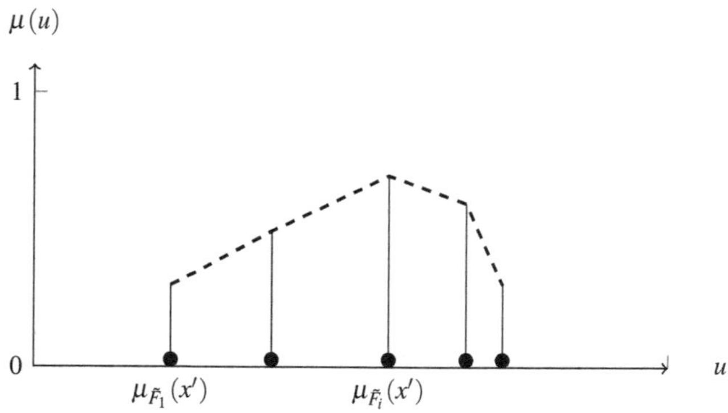

(b) Vertical slice over the FOU, evincing the variable secondary membership value of each embedded Fuzzy Set.

Fig. 2.10 Representation of a Type-2 Fuzzy Set

Type-2 FS representation can be constructed with the mean and standard deviations of scatter points obtained from surveys. More recently Wagner and Hagras [22] proposed a recursive algorithm to define an optimal approximation of the second degree membership function based on the collected data histogram for the cases of linguistic variables and noisy sensor measurements. The representation and manipulation algebras of Type-2 FSs are also non-closed problems, existing some recent proposals such as [23–25] that introduce simplifications in the Type-2 FSs's representation while maintaining the uncertainty in information representation over the inference stages.

2.5 Type-2 Fuzzy Logic Systems

The success of Type-1 FLSs naturally led to the development of FLSs based on Type-2 FSs. The structure of a Type-2 FLSs shares the same core components of its Type-1 counterpart, namely: a Fuzzifier, a Rule-Base, an Inference Engine and an Output Processor. While in Type-1 FLSs the final stage resumes to a defuzzification procedure, in the Type-2 case the Output Processor embraces an additional stage so that Type-2 FS is firstly converted into an equivalent Type-1 FS. This procedure is implemented by a Type-Reduction (TR) algorithm, which will be presented further in this document. The interdependency of the referred blocks is depicted in Fig. 2.11.

Type-2 FSs can be used either on antecedent, consequent or both levels of the Type-2 FLS, depending on whether is advantageous to account for uncertainties at the referred parts of the rule. As a consequence of its additional degrees of freedom, it has been argued that Type-2 FLSs have a great potential to produce better performing systems. The main reasons for this statement are the following:

- Given the fact that a Type-2 FS embeds itself a large number of Type-1 FS under the same label, it is possible to cover the same range of operation of a Type-1 FS with a smaller number of labels and rules, reducing the complexity of modeling, tuning and understanding a rule-base system comparatively to a similar performing Type-1 FS. This rule reduction capability is particularly advantageous in situations when the number of system inputs tend to increase, as it reduces the number of possible combinations of the linguistic labels that describe each input.
- In a Type-2 FLS, since each input and output is indirectly represented by a large number of Type-1 FSs, more complex input/output relationships that could not be obtained with in a Type-1 FLS can now be modeled without necessarily increase the number of rules [26].

While the additional degrees of freedom of Type-2 FSs revealed a considerable potential of supplanting conventional methodologies [23, 25], especially in complex non-linear modeling tasks, generally their use calls for a greater computational effort. Since Type-2 FS membership degrees are given as a function of a Type-1 FS, the formalisms of elementary fuzzy computations such as the union, intersection and

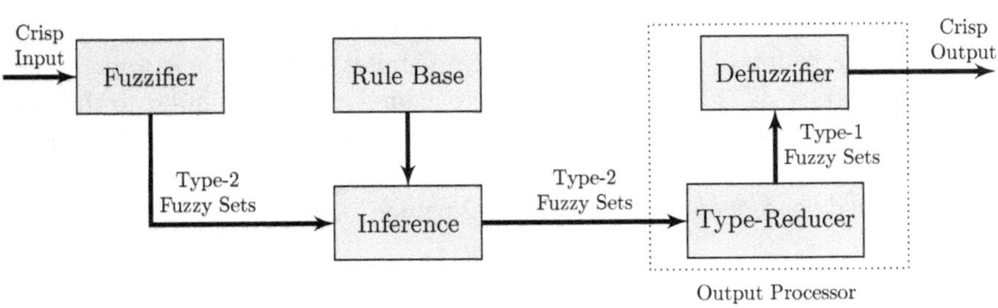

Fig. 2.11 Type-2 Fuzzy Logic System structure

complement are performed in a three-dimensional space, requiring far more complex procedures based on Zadeh's Extension Principle to implement such algebras. Moreover, the accuracy of the TR procedure depends on the number of discretization points of the FS input domain, which naturally is as better as many evaluation points are used - but the computational complexity also increases likewise. For that reason, a great amount of research in Type-2 Fuzzy Logic domains has been put towards developing more efficient representations in order to overcome this bottleneck and further extend its applicability to scenarios where the available computational resources are insufficient to cope with the time constants of the application scenarios.

In the past decade, Type-2 FSs' theory publication rate increased significantly, putting stronger efforts towards the reduction of its theoretical complexity and inherent computational effort - the two main problems that kept this uncertainty modeling tool away from real world applications until recent years. Shortly after its first publications, most of the authors focused on a simplified representation known as Interval Type-2 Fuzzy Sets (IT2FSs). In a IT2FS, the MFs uncertainty is restricted to the FOU and considers the third dimension of the Fuzzy Set as uniformly distributed with a membership value "1" [18]. As so, each primary membership is associated with the same third dimension, and each fuzzy set is characterized solely by its LMF and UMF. This concept is depicted in Fig. 2.12.

The Interval Type-2 FS can also be represented based on triangular, gaussian, trapezoidal or sigmoidal MFs. However, while one can define an arbitrary FOU as a piecewise function, the use of the referred classical shapes simplify further model adjustments in training procedures. For this reason, the literature mostly uses gaussian MFs since their FOU can be modeled by varying their mean and standard deviation, as follows:

$$\begin{aligned} \tilde{F}_i(c_i, \sigma_i, x) &= \exp\left[-\frac{1}{2}\left(\frac{x-c_i}{\sigma_i}\right)^2\right] \\ &= G(c_i, \sigma_i, x),\ \ \sigma_j^i \in [\sigma 1_i, \sigma 2_i]\ and\ c_i \in [c1_i, c2_i]\ . \end{aligned} \tag{2.11}$$

Figure 2.13 illustrates the resulting Type-2 FS by varying each one of the referred parameters individually.

The use of an interval based representation significantly reduced the complexity of all the calculations required in the FLSs and, for that reason, turned Interval Type-2 Fuzzy Logic Systems (IT2FLSs) feasible in practical scenarios. Despite the changes in the nature of the MFs, the basic principles of fuzzy logic remain valid and, consequently, IT2FSs' manipulation procedures are very similar to the ones already presented regarding its Type-1 counterpart. Following, a brief analysis of the Type-2 FLS based on the Mamdani inference will be performed assuming that both antecedent and consequent FSs are of Type-2 nature.

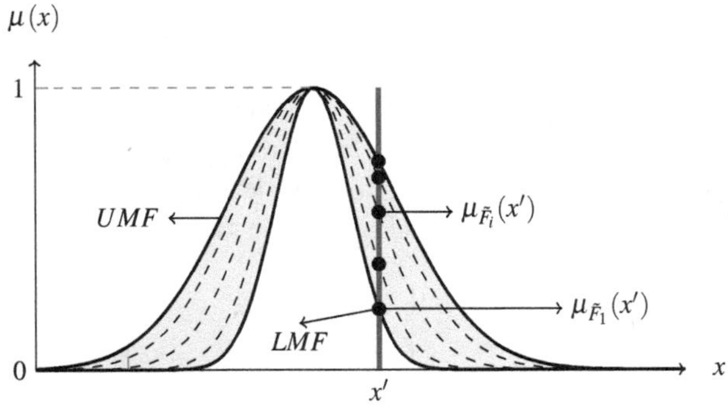

(a) FOU of an Interval Type-2 Fuzzy Set, evincing several embedded Fuzzy Sets.

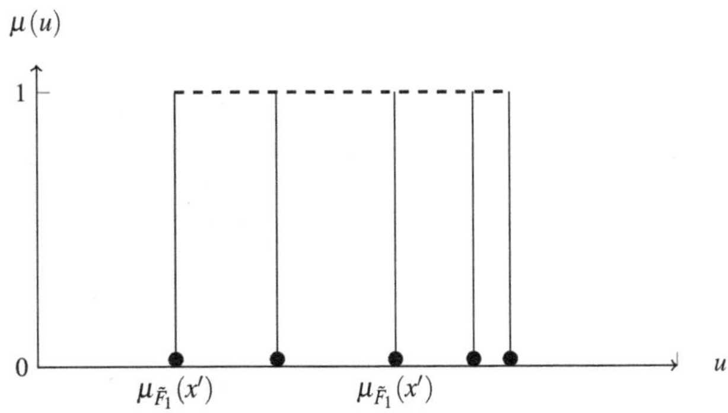

(b) Vertical slice over the FOU, evincing the constant secondary membership value of each embedded Fuzzy Set.

Fig. 2.12 Representation of an interval Type-2 Fuzzy Set

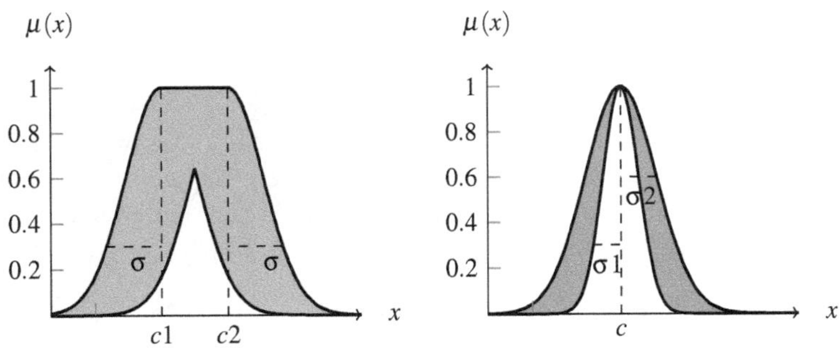

(a) Gaussian MF with uncertain mean. (b) Gaussian MF with uncertain standard deviation.

Fig. 2.13 Two possible representations of an interval Type-2 Fuzzy Set based on gaussian membership functions

2.5.1 Fuzzifier

Similarly to the Type-1 FLS, the simplest way of implementing the fuzzifier of a Type-2 FLS is to map a crisp input into a Singleton FS, as defined in Eq. (2.12). While information uncertainty is not explicitly considered in the fuzzification stage, it is indirectly accounted for in the rule's FSs representation

$$\mu_{\tilde{A}_x}(x) = \begin{cases} 1, \; x = x' \\ 0, \; \text{otherwise} \end{cases} \tag{2.12}$$

where x' is the system's input value.

2.5.2 Rule-Base

As a natural extension of Type-1 FLS, the Type-2 FLSs also synthesize their Rule-Base in a set of *If-Then* rules, establishing the relations between the system's input and output. Regardless the Fuzzy Sets nature, the procedure how rules are created remains the same. Therefore, a Type-2 FLS rule is represented as follows:

$$R^i : If \; x_1 \; is \; \tilde{F}_1^i \; and \; \cdots \; and \; x_j \; is \; \tilde{F}_j^i, \; Then \; y^i \; is \; \tilde{G}^i \; , \tag{2.13}$$

where R^i represents the ith fuzzy rule, $\tilde{F}_j^i$ and $\tilde{G}^i$ are linguistic terms characterized by Interval Type-2 FSs, $i = \{1, \cdots, M\}$ where M is the number of rules, $j = \{1, \cdots, N\}$ where N is the number of antecedents, x_j are the FLS inputs and y^i is the rule output.

2.5.3 Inference Engine

The main difference between a Type-1 FLS and a Type-2 FLS lies in their inference engine. In Sect. 2.2, it was concluded that the result of the jth input and corresponding antecedent operations in the ith rule yields a crisp number (μ_j^i) referred as membership degree. In an IT2FS the result of this operation is an interval given by $\tilde{\mu}_j^i$ as follows:

$$\tilde{\mu}_{F_j^i}(x_j) = \left[\underline{\mu}_{\tilde{F}_j^i}(x_j), \overline{\mu}_{\tilde{F}_j^i}(x_j) \right] \tag{2.14}$$

where x_j is the jth FLS system input.

Despite the apparent complexity of this result, an interval based representation allows the direct use of the basic fuzzy logic operations (union (*s-norm*), intersection (*t-norm*) and complement (*c-norm*)) as previously defined in Eqs. (2.5) and (2.6) by

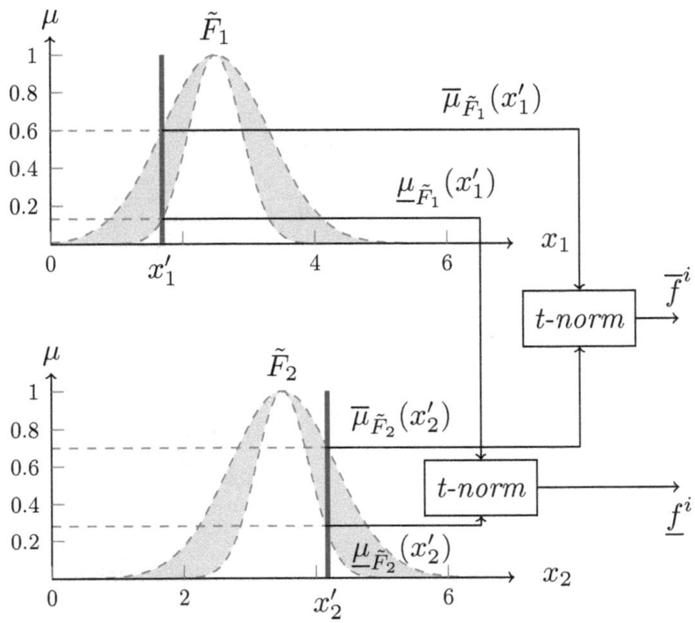

Fig. 2.14 Representation of the operation between singleton input and the antecedents of a Type-2 FLS using a *t-norm* operator (minimum or product)

considering the upper and lower bounds of the IT2FS separately. As so, the *t-norm* operator, which is used to perform the intersection of the antecedent FS is defined as:

$$\underline{f}^i = T_{j=1}^{N}\underline{\mu}_{\tilde{F}_j^i}(x_j) \qquad \overline{f}^i = T_{j=1}^{N}\overline{\mu}_{\tilde{F}_j^i}(x_j) \tag{2.15}$$

where T is a *t-norm* (product or minimum). The result of input and antecedent operations (for the minimum and product *t-norm*) is depicted in Fig. 2.14.

Similarly, the Mamdani implication methods (the Mamdani's minimum and product) can be directly used with IT2FS by applying the *t-norm* operator to the rule's firing level $\tilde{f}^i$ and the consequent $\tilde{G}_i$. This procedure is performed by considering the upper and lower bounds of $\tilde{f}^i$ and $\tilde{G}_i$ separately, as presented in Fig. 2.15 for the minimum and product *t-norms*.

The inference process yields a FS determined by the aggregation of the output of all the fired fuzzy sets. Similarly to the Type-1 FLS case, one can merge the contribution of each rule by finding the maximum value of the overlapped FSs, as depicted in Fig. 2.16. To obtain a crisp output after this procedure, one will have to apply a TR algorithm firstly, as will be discussed in the following subsection.

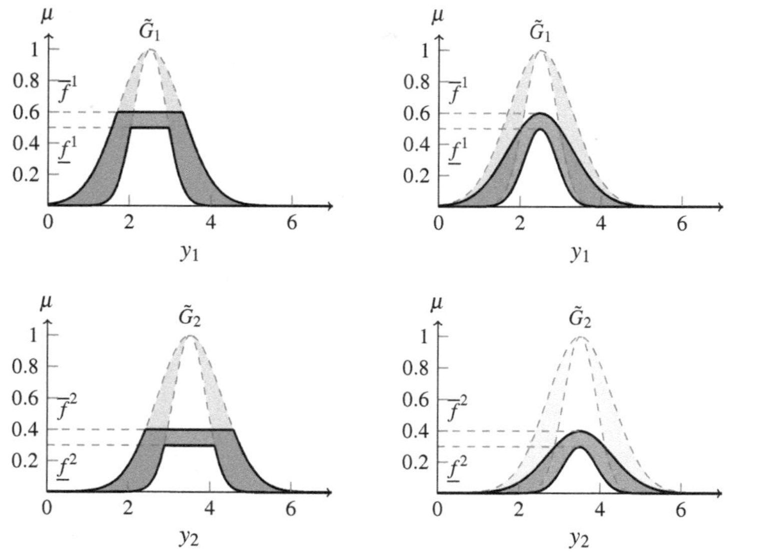

(a) Fired output sets using the minimum *t-norm*.

(b) Fired output sets using the product *t-norm*.

Fig. 2.15 Mamdani inference operations using Type-2 FSs

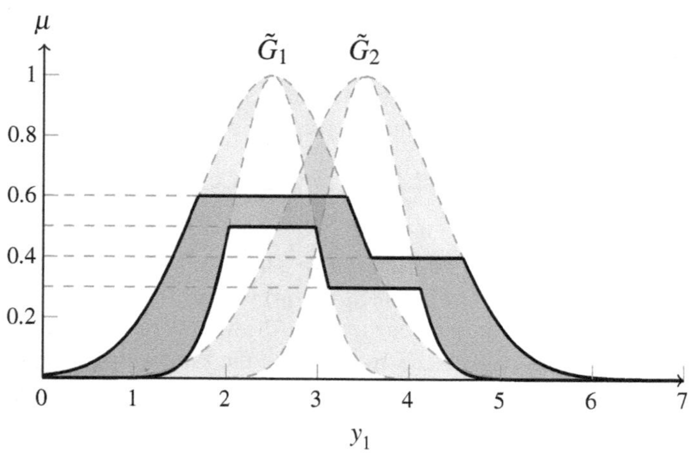

Fig. 2.16 Type-2 Fuzzy Sets fired consequents' aggregation procedure, after using the Mamdani minimum implication

2.5.4 Type-Reduction

In order to develop practical applications based on the Type-2 FL, it becomes necessary to obtain a crisp value from the combination of all fired FS. To accomplish this goal, it is a prerequisit to obtain the centroid of a Type-2 FS, represented as an interval often referred to as type-reduced set. The Karnik-Mendel (KM) algorithm [27], which can be seen as an extension of Type-1 defuzzification procedure, is currently the most accurate TR method found in literature. Though, given its iterative nature, it is the most complex stage of the fuzzy inference process, requiring extensive calculations even when the simpler IT2FSs are used.

2.5.4.1 Karnik-Mendel Type-Reduction

The KM algorithm is an iterative process which allows one to obtain an interval of uncertainty given by $[y_l, y_r]$ for the centroid of an Interval Type-2 FS. Similarly to the Type-1 FSs defuzzification case, Karnik and Mendel [13] also proposed several methods to perform the Type-2 FSs' TR based on approaches of the Type-1 FLS defuzzification procedures, namely: Height and Modified Height TR, Centroid TR and Center-of-Sets TR. The choice of the defuzzification method has significant implications in the result's quality. The Height and Modified Height are the less complex ones to implement. However, it is known that when a single rule is triggered, these methods may return inconsistent results [13]. The Centroid one requires a large amount of calculations because, for each new system input, it has firstly to merge the consequent part FS of every rule and only then obtain the centroid of the resulting FS. Finally, the Center-of-Sets TR is usually the employed method since it performs a smaller amount of operations when compared with the Centroid one. Its efficiency is due to the *a priori* computation of each consequent FS's centroid. Since the FS's centroid value is independent from the system's input variables, this result can be used as a constant in the Center-of-Sets TR. Therefore, the only procedure that has to be performed after each new input into the system is a weighted average of the stored centroids according to a combination of the upper and lower firing levels of each rule. Since the Center-of-Sets TR inevitably requires one to compute the centroid of each consequent Type-2 FS once, the Centroid TR will be hereby presented.

Similarly to the Centroid defuzzification procedure, the Centroid TR starts by obtaining K samples from a Type-2 FS. Since the FOU of a Type-2 FS embeds several Type-1 FS, to perform the TR one has firstly to obtain two Type-1 FS whose centroid best approximates the upper and lower bounds of the Type-2 FS centroid. Considering for example the $\tilde{G}_{out}$ FS, the procedure starts by using its sampled upper and lower bounds to find the optimal values for the switching points [L, R], as depicted in Fig. 2.17.

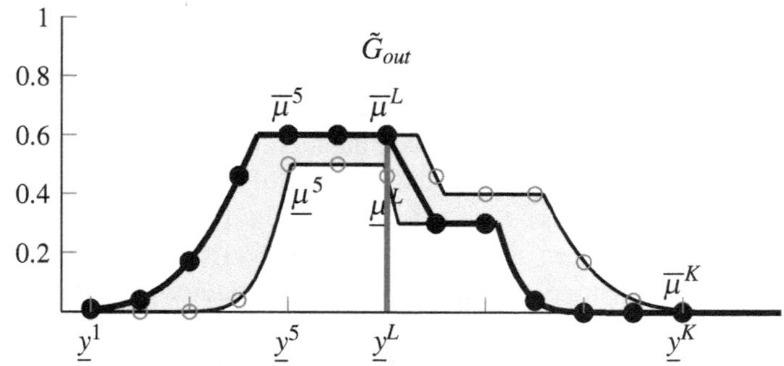

(a) Computing y_l: Switching from the upper bounds of the firing intervals to the lower bounds.

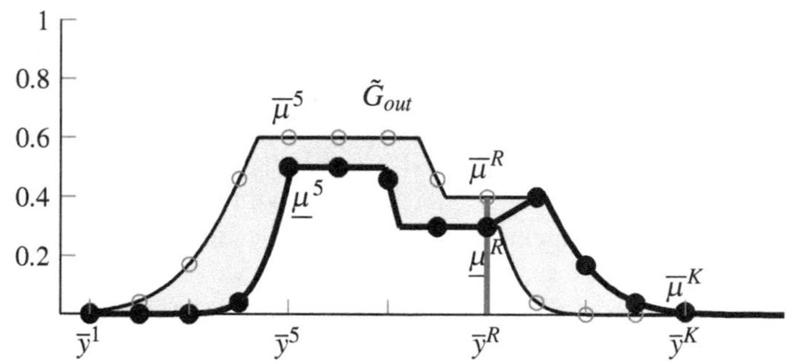

(b) Computing y_r: Switching from the lower bounds of the firing intervals to the upper bounds.

Fig. 2.17 Switching points in computing y_l and y_r

The candidate points are obtained as follows in Eqs. (2.16) and (2.17).

$$y_l(k) = \frac{\sum_{i=1}^{k} \underline{y}^i \overline{\mu}^i_{\tilde{G}_{out}} + \sum_{i=k+1}^{K} \underline{y}^i \underline{\mu}^i_{\tilde{G}_{out}}}{\sum_{i=1}^{k} \overline{\mu}^i_{\tilde{G}_{out}} + \sum_{i=k+1}^{K} \underline{\mu}^i_{\tilde{G}_{out}}}, \tag{2.16}$$

$$y_r(k) = \frac{\sum_{i=1}^{k} \overline{y}^i \underline{\mu}^i_{\tilde{G}_{out}} + \sum_{i=k+1}^{K} \overline{y}^i \overline{\mu}^i_{\tilde{G}_{out}}}{\sum_{i=1}^{k} \underline{\mu}^i_{\tilde{G}_{out}} + \sum_{i=k+1}^{K} \overline{\mu}^i_{\tilde{G}_{out}}}, \tag{2.17}$$

where k is an integer in $[1, K - 1]$ interval, and K represents the number of discretization points. Then, the optimal interval bounds can be obtained by y_l and y_r, as follows:

$$y_l = \min_{k \in [1, M-1]} y_l(k) \equiv y(L) \equiv \frac{\sum_{i=1}^{L} \underline{y}^i \overline{\mu}^i_{\tilde{G}_{out}} + \sum_{i=L+1}^{K} \underline{y}^i \underline{\mu}^i_{\tilde{G}_{out}}}{\sum_{i=1}^{L} \overline{\mu}^i_{\tilde{G}_{out}} + \sum_{i=L+1}^{K} \underline{\mu}^i_{\tilde{G}_{out}}}, \tag{2.18}$$

$$y_r = \max_{k \in [1, M-1]} y_r(k) \equiv y(R) \equiv \frac{\sum_{i=1}^{R} \overline{y}^i \underline{\mu}^i_{\tilde{G}_{out}} + \sum_{i=R+1}^{K} \overline{y}^i \overline{\mu}^i_{\tilde{G}_{out}}}{\sum_{i=1}^{R} \underline{\mu}^i_{\tilde{G}_{out}} + \sum_{i=R+1}^{K} \overline{\mu}^i_{\tilde{G}_{out}}}, \tag{2.19}$$

where L and R are switch points satisfying

$$y^L \leq y_l < y^{L+1}, \tag{2.20}$$

$$y^R \leq y_r < y^{R+1}. \tag{2.21}$$

The choice of whether we start from the upper or lower firing levels when finding the left and right bounds of each switching point has a very simple explanation. Take y_l as an example: y_l has to be the minimum value of the FLS output. Since $\underline{y}^i$ is ordered ascendantly along the horizontal axis of Fig. 2.12, a large weight (upper bound of the firing interval) should be chosen in the left of the switch point and a small weight (lower bound of the firing interval) for its right side. As finding all the centroid $[y_l, y_r]$ candidates is a computationally inefficient approach, an iterative procedure to find the optimal switching points is presented in Table 2.1.

Despite the improvements brought by Interval Type-2 FS representations, the KM algorithm still requires a large number of iterations to find the optimal type-reduced FS. Therefore, several enhancements and simplifications were proposed in the recent years for the sake of reducing its computational footprint.

2.5.4.2 Optimized Type-Reduction Algorithms

With the development of simpler and alternative algorithms, Type-2 FL definitely gathered the attention of a broader number of researchers, having a direct impact in an increasing number of applications in domains such as modeling, control and classification and pattern recognition observed in recent years.

Table 2.1 Iterative Karnik-Mendel algorithm

Step	For computing y_l	For computing y_r
1.	Initialize $\mu^i_{\tilde{G}_{out}} = \frac{\underline{\mu}^i_{\tilde{G}_{out}} + \overline{\mu}^i_{\tilde{G}_{out}}}{2}$ and compute $y = \frac{\sum_{i=1}^{M} \underline{y}^i \mu^i_{\tilde{G}_{out}}}{\sum_{i=1}^{M} \mu^i_{\tilde{G}_{out}}}$	Initialize $\mu^i_{\tilde{G}_{out}} = \frac{\underline{\mu}^i_{\tilde{G}_{out}} + \overline{\mu}^i_{\tilde{G}_{out}}}{2}$ and compute $y = \frac{\sum_{i=1}^{M} \overline{y}^i \mu^i_{\tilde{G}_{out}}}{\sum_{i=1}^{M} \mu^i_{\tilde{G}_{out}}}$
2.	Find $l \in [1, M-1]$ such that $\underline{y}^l < y < \underline{y}^{l+1}$	Find $r \in [1, M-1]$ such that $\overline{y}^r < y < \overline{y}^{r+1}$
3.	Set $\mu^i_{\tilde{G}_{out}} = \begin{cases} \overline{\mu}^i_{\tilde{G}_{out}}, & n \leq l \\ \underline{\mu}^i_{\tilde{G}_{out}}, & n > l \end{cases}$ and compute $y' = \frac{\sum_{i=1}^{M} \underline{y}^i \mu^i_{\tilde{G}_{out}}}{\sum_{i=1}^{M} \mu^i_{\tilde{G}_{out}}}$	Set $\mu^i_{\tilde{G}_{out}} = \begin{cases} \underline{\mu}^i_{\tilde{G}_{out}}, & n \leq r \\ \overline{\mu}^i_{\tilde{G}_{out}}, & n > r \end{cases}$ and compute $y' = \frac{\sum_{i=1}^{M} \overline{y}^i \mu^i_{\tilde{G}_{out}}}{\sum_{i=1}^{M} \mu^i_{\tilde{G}_{out}}}$
4.	If $y' = y$, stop and set $y_l = y$ and $L = l$; otherwise, set $y = y'$; and go to step 2	If $y' = y$, stop and set $y_r = y$ and $R = r$; otherwise, set $y = y'$; and go to step 2

The Type-Reduction methods found in literature can be grouped into two main categories:

- Enhancements to the KM algorithm, that improve directly the original formulation of the KM by choosing a better initialization and termination conditions, to reduce the number of iterations and to optimize the computing technique speeding up each iteration of the TR process.
- Alternative TR algorithms, which unlike the iterative KM algorithms, are mostly presented in a closed-form representation and provide faster results than the KM method.

In [28] a thorough analysis about the current TR algorithms' state-of-the-art is done and it was observed that enhanced versions of the KM algorithm are, in general, faster than its original formulation. Yet, the gains may vary with the size of the FLS rule base. From the presented approaches, the Enhanced Opposite Direction Searching Algorithm (EODS) [29] shown itself as the fastest one achieving gains up to 70% (relative to the original KM) when the FLS has less than 100 rules—as is used in most part of non-linear processes' modeling applications. It is important to highlight that, despite their algorithmic differences, enhanced versions of the KM algorithms give exactly the same outputs as its original formulation.

While KM algorithms have been widely adopted, some closed-form methods that bypass this TR procedure have been proposed. However, since their methodologies may be significantly different than the original KM algorithms their outputs may also be quite different. Therefore, a compromise between accuracy and complexity of the method may be necessary. For example, Wu and Tan [30] introduced a method which eliminates TR by defining a collection of Type-1 FS embedded by the footprint of uncertainty. Alternatively, Wu-Mendel Uncertainty Bound method [31] directly uses the uncertainty bounds of the FS and was shown to be the closed-form method giving the closest approximation to the KM and presented an execution turnaround time very close to a similar sized Type-1 FLS. Despite their good performance, a closed-form approach is not adopted in this work since the best performing ones hinder the decomposition of the model output as sum of locally linear models. The capability of decomposing the system in such way is of great advantage to the present work as it allows an efficient implementation of the online training procedures and the synthesis of the control law based on the Generalized Predictive Control theory.

2.5.5 *Defuzzifier*

After applying one of the possible TR methods, the obtained Interval Fuzzy Set still has to be converted into a crisp number so it becomes suited to the most part of the FLS application scenarios. Anyway, this procedure is fairly straightforward, and the defuzzified value obtained by simply computing the average of the interval's left and right endpoints as:

$$y_{out} = \frac{y_r + y_l}{2} \,. \tag{2.22}$$

2.6 Comparative Analysis

The noise reduction properties of Type-2 Fuzzy MFs have been several times pointed in literature as one of its main advantages when compared to its Type-1 counterparts. To study the influence of the antecedent part membership functions' FOU width in the rule activation level when the FLS's inputs are corrupted by a disturbance, a comparative analysis will be performed by probing its input domain with different magnitude noise levels. This comparison will be based on Gaussian-shaped FS with uncertain mean (as presented in Fig. 2.13, where its FOU is bounded by the upper and lower MFs as defined by the equations:

$$\underline{\mu}_{\tilde{F}}(x) = \begin{cases} G(c2, \sigma, x), & x < \frac{c1+c2}{2} \\ \\ G(c1, \sigma, x), & x \geq \frac{c1+c2}{2} \end{cases}, \tag{2.23}$$

$$\overline{\mu}_{\tilde{F}}(x) = \begin{cases} G(c1, \sigma, x), & x < c1 \\ 1, & c1 \leq x \leq c2 \\ G(c2, \sigma, x), & x > c2 \end{cases}. \tag{2.24}$$

In a similar approach as presented in [32], this procedure will be based on a simple FLS comprising a single input and two rules defined by two overlapping MFs ($\tilde{F}_1$ and $\tilde{F}_2$) such that, for a certain input value x, the following conditions are satisfied:

$$\overline{\mu}_2 = 1 - \underline{\mu}_1 , \tag{2.25a}$$

$$\underline{\mu}_2 = 1 - \overline{\mu}_1 , \tag{2.25b}$$

where $\overline{\mu}_i$ and $\underline{\mu}_i$ are the upper and lower firing levels of $\tilde{F}_i(x)$, respectively.

To simplify this evaluation, it is considered that the output is given by Nie Tan closed form Type-Reduction [33],

$$y = \frac{\sum_{i=1}^{M} (\underline{f}^i + \overline{f}^i) y^i}{\sum_{i=1}^{M} (\underline{f}^i + \overline{f}^i)} \tag{2.26}$$

where y_i is the output of each rule and $\underline{f}^i$ and $\overline{f}^i$ are equivalent to $\overline{\mu}_i$ and $\underline{\mu}_i$, respectively, since the system solely has one antecedent. Therefore, based on Eqs. (2.25) and (2.26), the contribution of each rule to the model output is weighted by

$$\begin{aligned} r_i(x) &= \frac{\underline{f}^i + \overline{f}^i}{\sum_{k=1}^{M} (\underline{f}^k + \overline{f}^k)} \\ &= \frac{\underline{\mu}^i(x) + \overline{\mu}^i(x)}{2} . \end{aligned} \tag{2.27}$$

If the system's input is corrupted by a gaussian noise of magnitude n, then the firing strength becomes:

$$r_i(x+n) = \frac{\underline{\mu}_i(x+n) + \overline{\mu}_i(x+n)}{2} . \tag{2.28}$$

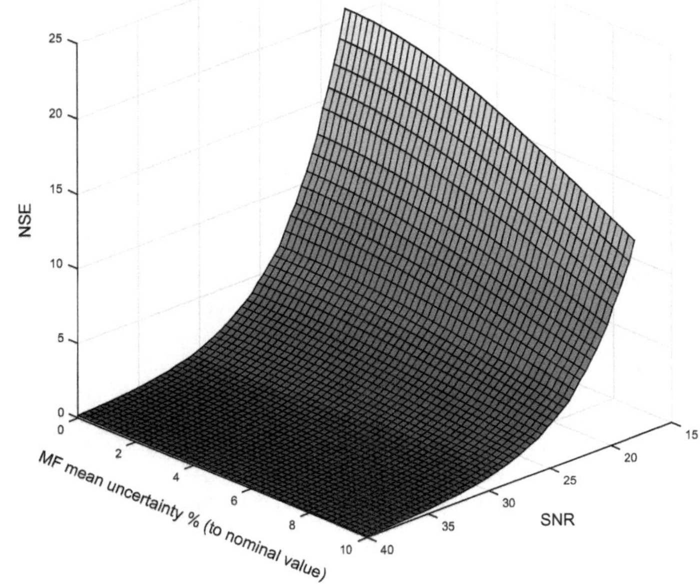

Fig. 2.18 Surface of the NSE of the evaluated system depending on the input noise level and the antecedent parameter's FOU width

Based on Eqs. (2.27) and (2.28), one can evaluate the relationship between the firing level distortion caused by the input noise and the FOU width for a single rule by obtaining the Normalized Squared Error (NSE) as

$$NSE = \int_{n=-n_1}^{n_1} \int_{x=x_1}^{x_2} \left[\frac{r_i(x) - r_i(x+n)}{r_i(x)} \right]^2 dxdn, \qquad for\ n \in \mathbb{R} \quad x \in \mathbb{R}\,. \tag{2.29}$$

The solution of Eq. (2.29) is obtained numerically by varying in the same proportion the values c_1 and c_2 relatively to a Type-1 FS initially centred in c and then corrupting the input variable with a noise signal with different Signal-to-Noise Ratios (SNRs), defined relatively to the maximum input value of the fuzzy set's domain. The maximum uncertainty percentage relatively to the FS upper and lower MF's center is limited to 10%, as with higher values a great portion of the input space a large portion of the upper and lower membership degrees become closer to one and zero respectively, thus not providing a desirable input space partition for a TS system. Figure 2.18 presents the dependency of the NSE on the noise level and the uncertainty ratio, and the results obtained by averaging 50 trials for every parameters' combination.

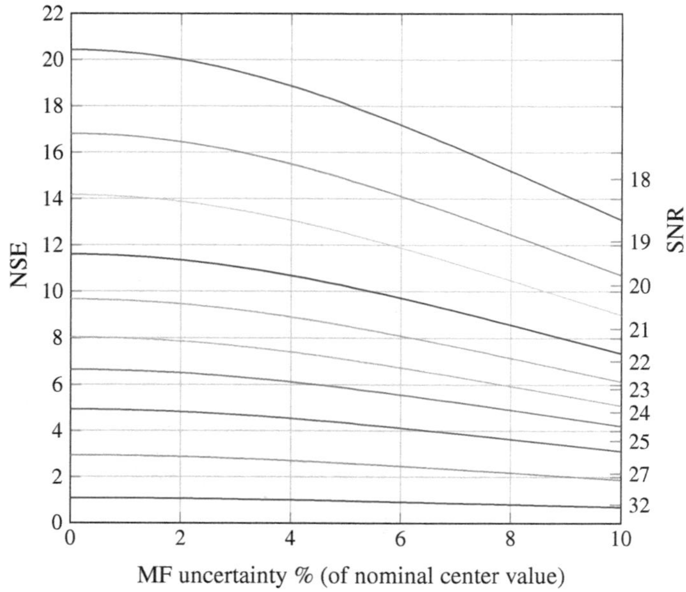

Fig. 2.19 Dependency of the firing degree NSE on the membership function's center uncertainty ratio and the noise level corrupting the input signals

Table 2.2 Distortion level for different SNR considering 10% uncertainty over the membership function's center.

SNR	18	19	20	21	22	23	24	25	27	32
NSE	13.11	10.64	8.98	7.35	6.13	5.09	4.21	3.13	1.88	0.70
σ	0.98	0.81	0.72	0.47	0.46	0.39	0.38	0.24	0.13	0.05

For a better comparison between the different scenarios, a bi-dimensional projection perspective of the previous surface is presented in Fig. 2.19 and the results relative to the 10% uncertainty level summarized in Table 2.2.

From the presented results two main conclusions can be obtained regarding the noise properties of Type-2 FSs:

- For a given SNR, increasing its FOU reduces the distortion observed at each rule firing level when comparing to the Type-1 counterpart which served as starting point (when a 0% uncertainty factor is considered in the MF's center);
- For reduced noise levels, the use of Type-2 FS at the antecedent part of the rule base does not bring significant improvements comparing to its Type-1 counterpart, as is evinced by the flatter region observed in Fig. 2.19 as SNR is increased.

2.7 Conclusions

This chapter presented the fundamental theory of Type-1 Fuzzy Logic Systems and how its extension to the Type-2 Fuzzy Logic formalisms can be performed. In the recent years, Type-2 Fuzzy Logic has been acclaimed as a significant improvement over the fundamental Fuzzy Logic theory. Despite the lack of an irrefutable theoretical proof of it, the fact is that most recent publications present practical applications where the use of Type-2 Fuzzy Logic is shown to be advantageous, mostly in scenarios where uncertain information representations are manipulated. The evaluative scenario concluding this chapter points also in that direction. Therefore, the concepts here introduced will be further developed in the succeeding chapters, by integrating them with more flexible structures in terms of learning capabilities.

References

1. Albus, J.: A therory of cerebellar function. Math. Biosci. **10**, 25–61 (1971)
2. Newell, A., Simon, H.: Human Problem Solving. Prentice-Hall (1972)
3. Zadeh, L.: Fuzzy sets. Inf. Control **8**, 338–353 (1965)
4. Mamdani, E.: Applications of fuzzy algorithms for control of a simple dynamic plant. Proc. IEEE **121**, 1585–1588 (1974)
5. Verbruggen H., Babuska, R.: Fuzzy Logic Control: Advances in Applications. World Scientific (1999)
6. Yasunobu, S., Miyamoto, S., Ihara, S.: Train automatic operation system by fuzzy theory. In: Proceedings of 20th SICE, pp. 467–468 (1981)
7. Yamakawa, T.: Stabilization of an inverted pendulum by a high-speed fuzzy logic controller hardware system. Fuzzy Sets Syst. **32**(2), 161–180. Elsevier (1989)
8. Rumerman, J.: NASA Launch Systems, Space Transportation/Human Spaceflight, and Space Science 1989–1998, NASA Historical Data Book, vol. VII, The NASA History Series, Volume VII, (2009)
9. Mamdani, E.: Applications of fuzzy logic to approximate reasoning using linguistic systems. IEEE Trans. Syst. Man Cybern. **26**(12), 1182–1191 (1977)
10. Takagi, T., Sugeno, M.: Fuzzy identification of systems and its applications to modeling and control. IEEE Trans. Syst. Man Cybern. **15**(1), 116–132 (1985)
11. Kosko, B.: Fuzzy systems as universal approximators. IEEE Trans. Comput. **43**(11), 1329–1333 (1994)
12. Delgado, M., Duarte, O., Requena, I.: Arithmetic approach for the computing with words paradigm. Int. J. Intell. Syst. **21**, 121–142. Wiley (2006)
13. Mendel, J.: Uncertain Rule-Based Fuzzy Logic Systems: Introduction and New Directions. Prentice-Hall (2001)
14. Passino, K., Yurkovich, S.: Fuzzy Control. Addison-Wesley (1998)
15. Mendel, J., Zadeh, L.: Trillas, E., Yager, R., Lawry, J., Hagras, H., Guadarrama, S.: What computing with words means to me. IEEE Comput. Intell. Mag. (2010)
16. Zadeh, L.: The concept of linguistic variable and its applications to approximate reasoning. Inf. Sci., Part I–III, pp. 199–249, pp. 301–357, pp. 43–80 (1975)
17. Karnik, N., Mendel, J., Liang, Q.: Type-2 fuzzy logic systems. IEEE Trans. Fuzzy Syst. **7**(6), 643–658 (1999)
18. John, R., Coupland, S.: Type 2 fuzzy logic: a historical view. IEEE Comput. Intell. Mag. (2007)

19. Mizumoto, M., Tanaka, K.: Fuzzy sets of type-2 under algebraic product and algebraic sum. Fuzzy Sets Syst. **5**, 277–290 (1981)
20. Dubois, D., Prade, H.: Fuzzy Sets and Systems: Theory and Applications. Academic Press (1982)
21. Turksen, I., Norwich, A.: Measurement of fuzziness, measurement of fuzziness. In: Proceedings of the International Conference on Policy Analysis and Information Systems, pp. 745–754 (1981)
22. Wagner, C., Hagras, H.: Novel methods for the design of general type-2 fuzzy sets based on device characteristics and linguistic labels surveys. In: Proceedings of the 2009 International Fuzzy Systems Association World Congress and the 2009 European Society for Fuzzy Logic and Technology Conference, pp. 537–543 (2009)
23. Wagner, C., Hagras, H.: Towards general type-2 fuzzy logic systems based on zSlices. IEEE Trans. Fuzzy Syst. **18**(4) (2010)
24. Coupland, S., John, R.: Geometric type-1 and type-2 fuzzy logic. IEEE Trans. Fuzzy Syst. **15**, 3–15 (2007)
25. Wagner, C., Miller, S., Garibaldi, J., Anderson, D.: From interval-valued data to general type-2 fuzzy sets. IEEE Trans. Fuzzy Syst. **23**(2), 248–269 (2015)
26. Wu, D.; Tan, W.: Type-2 FLC modeling capability analysis. In: Proceeding of the 2005 IEEE International Conference on Fuzzy Systems, pp. 242–247 (2005)
27. Karnik, N., Mendel, J.: Centroid of a type-2 fuzzy set. Inf. Sci. **132**, 195–220 (2001)
28. Wu, D.: Approaches for reducing the computational cost of interval type-2 fuzzy logic systems: overview and comparisons. IEEE Trans. Fuzzy Syst. **21**(1) (2013)
29. Hu, H., Wang, Y., Cai, Y.: Advantages of enhanced opposite direction searching algorithms for computing the centroid of an interval type-2 fuzzy set. Asian J. Control **14**(6), 1–9 (2012)
30. Wu, D.; Tan, W.: Computationally efficient type-reduction strategies for a type-2 fuzzy logic controller. In: Proceedings of IEEE International Conference in Fuzzy Systems, pp. 353–358 (2005)
31. Wu, D., Mendel, J.: Uncertainty bounds and their use in the design of interval type-2 fuzzy logic systems. IEEE Trans. Fuzzy Syst. **10**(5), 622–639 (2002)
32. Kayacan, E.: interval type-2 fuzzy logic systems: theory and design, Ph.D. Thesis, Bogaziçi University, Istanbul, Turkey (2011)
33. Nie, M.; Tan, W.: Towards an efficient type-reduction method for interval type-2 fuzzy logic systems. In: Proceedings of IEEE International Conference on Fuzzy Systems, pp. 1425–1432 (2008)

Chapter 3
Takagi-Sugeno Fuzzy Logic Systems

3.1 Introduction

The achievements obtained by Fuzzy Logic undoubtedly changed the way expert information is represented, manipulated, and interpreted in computational systems. Nevertheless, the initialization of Mamdani FLSs' main parameters, namely its membership functions and their interdependency relations, is a process that depends on the knowledge of an expert (which may be subjective and is ultimately limited by its know-how). Takagi and Sugeno [1] were among the first researchers who recognized that FLSs could be further enhanced with autonomous learning techniques. Together, they proposed a new structure for the consequent part of the rules, introducing also methodologies to autonomously create and improve the FLSs' performance. Their method uses heuristic and non-linear optimization algorithms for the antecedent part of the rule-base and a Kalman Filter for the consequent one. It is however for their innovative FLS's structure supporting their work that Takagi and Sugeno are nowadays known in FLSs' literature (effectively coining the concept of Takagi-Sugeno FLSs), serving their work as the stepping stone for many successful research topics.

In what concerns to learning ability, Artificial Neural Networks(ANNs) stand in the exact opposite side of traditional FLSs. ANNs can approximate any arbitrary function representing a system's input/output behavior by means of a network of several activation functions, with parameters that can be autonomously tuned based on simple concepts as the output error back-propagation. It is a problem, though, that the knowledge of these systems is stored in an opaque way for the system's designer since the learning results are stored in a large set of parameter values with hardly any interpretable features. To overcome such limitation and improve FLSs adaptability, different structures inspired by Multi-Layer Neural Networks have been presented over the last years. These hybrid architectures, referred to as Neuro-Fuzzy Systems [2], reveal themselves as an approach that benefits from the readability of a fuzzy rule and the learning ability of ANNs. Among many Neuro-Fuzzy architectures, the most referred ones are the Fuzzy Adaptive Learning Control Network (FALCON) [3], the Generalized Approximate Reasoning based Intelligence Control

R. Antão, *Type-2 Fuzzy Logic*, Nonlinear Physical Science,
DOI 10.1007/978-981-10-4633-9_3

(GARIC) [4], Neural Fuzzy Controller (NEFCON) [5], the Self Constructing Neural Fuzzy Inference Network (SONFIN) [6] and ultimately the Adaptive network based Fuzzy Inference System (ANFIS) [7].

This multitude of Neuro Fuzzy Systems(NFSs)'s implementations is in its essence similar, but present some fundamental differences [8]: while some use ANNs as a pre or post-processing stage for the FLSs, other focus on reorganizing well known structures such as the Mamdani FLS or the Takagi-Sugeno FLS ones into an equivalent Multi-Layer Neural Network. Some authors take the system's learning capabilities further ahead, proposing algorithms to develop the model in a completely autonomous approach, providing system-wide adaption mechanisms to optimize their structure. However, most of them focus on the parameter level adaption, leaving the structure problem up to an application expert analysis (effectively making use of the Fuzzy Systems intelligibility).

The ANFIS architecture is one of the most successful Neuro-Fuzzy systems' implementations due to its functional equivalence to the TS FLSs, providing a simple methodology to convert *If-Then* rules into an adaptive Radial Basis Function Network (RBFN). Due to the typical TS FLS formulation, where each rule's output is given by a function of its input variables, the conversion of the inference and aggregation procedures is fairly straightforward. A similar procedure could also be performed according to the Mamdani type of FLS, but its applicability it is restricted to very specific types of defuzzification procedures (Center-of-Sets) for it to become a computationally efficient alternative. As will be clear further in this book, the balance between computational speed and methodology accuracy are major concerns when a FLS is used as a model in real-time systems, thus making the TS systems better candidates than the Mamdani ones to accomplish such task.

Since the structures of TS FLSs and ANFIS are deeply related, this chapter will begin by presenting the former one, starting with the Type-1 FS which will be then extended to the several possible architectures based on Type-2 ones. Then, the procedures employed for training of the TS FLS will be presented.

3.2 Type-1 Takagi-Sugeno Fuzzy Logic Systems

Takagi-Sugeno FLS are a type of Fuzzy Systems' which, along with the Mamdani one, became the *de facto standards* in fuzzy modeling and control applications. Similarly to the Mamdani FLSs, the Takagi and Sugeno [1] ones establish an input-output relation based on a set of *If-Then* rules. In the latter case, while the system's input space is partitioned by Type-1 FS, the consequent part of each rule is usually given by a first order polynomial. Even though it is possible to use higher-order polynomials, first-order ones are widely preferred due to their closeness with linear modeling techniques [9]. Equation (3.1) presents the structure of a first-order Type-1 TS model rule,

$$\begin{aligned} R^i : \quad & If\ x_1\ is\ F_1^i\ and\ \cdots\ and\ x_j\ is\ F_j^i \\ & Then\ y^i = c_1^i x_1 + \cdots + c_j^i x_j\ , \end{aligned} \tag{3.1}$$

where R^i represents the ith fuzzy rule, F^i_j are linguistic terms characterized by Type-1 FS, c^i_j are the consequent polynomial parameters, $i = \{1, \cdots, M\}$ where M is the number of fuzzy rules, $j = \{1, \cdots, N\}$ where N is the number of antecedents, x_j are the fuzzy system inputs and y^i is the rule output.

The algebraic nature of each rule's consequent part inherently implements defuzzification mechanism in TS FLSs, not requiring any further steps to convert a fired output Type-1 FS into an equivalent crisp value, as opposed to the Mamdani case. Consequently, the global output of a Type-1 TS FLS is obtained straightforwardly using the Center-of-Sets defuzzification, which is no more than a weighted average of the output of the M rules according to their firing level, as follows in Eq. (3.2):

$$y(\mathbf{x}) = \frac{\sum_{i=1}^{M} f^i y^i}{\sum_{i=1}^{M} f^i} = \frac{\sum_{i=1}^{M} f^i (c^i_1 x_1 + c^i_2 x_2 + \cdots + c^i_N x_N)}{\sum_{i=1}^{M} f^i} , \tag{3.2}$$

where f^i is the rule's firing level, defined as:

$$f^i = T^N_{k=1} \mu_{F^i_k}(x_k) , \tag{3.3}$$

and $T^N_{k=1}$ denotes a *t-norm*, a operator which merges the firing levels of each rule's antecedent. Similarly to the aggregation procedures presented on the previous chapter, the minimum and product operators are usually employed. The latter the most commonly used and also adopted in this work. As so, f^i becomes:

$$f^i(\mathbf{x}) = \prod_{k=1}^{N} \mu_{F^i_k}(x_k) . \tag{3.4}$$

In Fig. 3.1 the main concepts supporting the TS FLS inference mechanism are depicted.

While providing a relatively simple model structure and maintaining an important level of decipherability, the Takagi–Sugeno FLSs also offer an efficient and accurate way of modeling non-linear behaviors. The antecedent part of the *If-Then* rules allows one to partition a system's input space using several input/output linear functions, which are valid approximations of the global non-linear system under different operating regions. Considering a particular local model output, given by $y^i(\mathbf{x})$ and defined for an operation point in the vicinity of $\mathbf{x}$, its validity for the current operating regime (given by the input vector $\mathbf{x} = [x_1, x_2, \cdots, x_N]$) is as high as its firing level (f^i) is closer to unity, and consequently lower when the local approximation is no longer valid. Figure 3.2 presents a simple case where such approach can be used to model a non-linear input/output relationship.

The overall non-linear behavior of the system can be obtained by a smooth interpolation of simpler local linear subsystems which are transparent up to a certain degree.

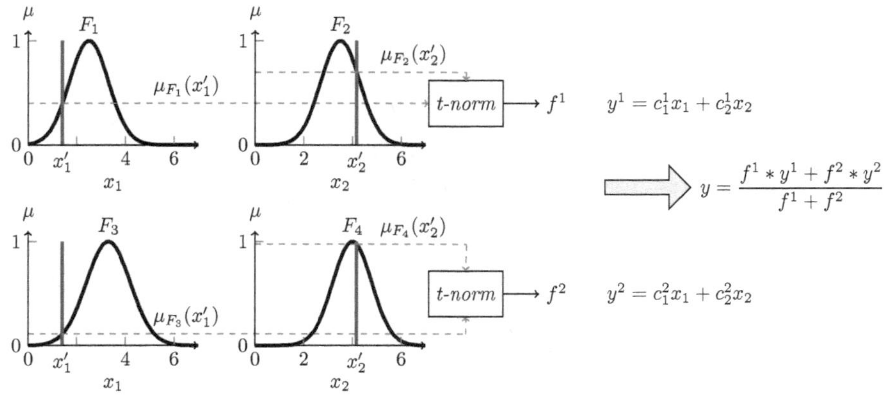

Fig. 3.1 Example of a two rules—two antecedent Type-1 Takagi-Sugeno FLS

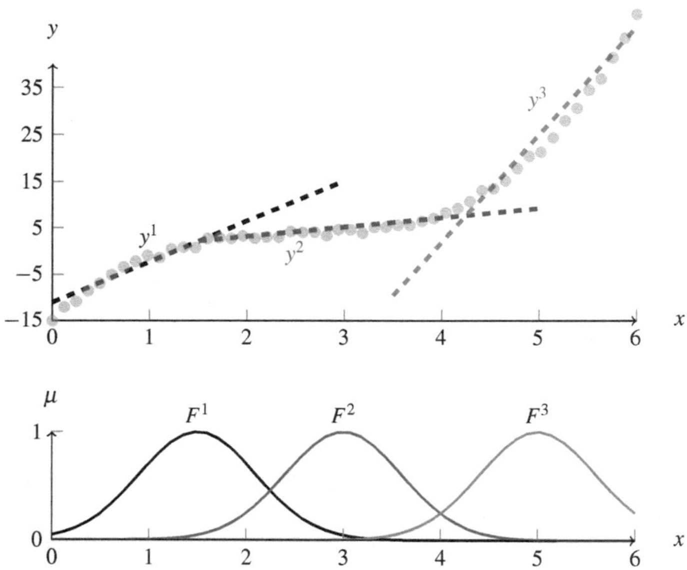

Fig. 3.2 Example of a three rule ($M = 3$) partition with a single input ($N = 1$) TS FLS to model a non-linear function

Such feature overcomes the limitations of some black-box non-linear modeling methods such as Neural-Networks or Volterra series, where the model dimensionality can increase significantly and the relationships between the system variables becomes intractable. Demonstrating the theoretical and practical utility of TS FLSs, it has also been proved that they are also universal approximators [7], attesting their capability of approximate any reasonable function with subjective accuracy depending on the number of rules and the training level.

3.3 Type-2 Takagi-Sugeno Fuzzy Logic Systems

Inspired by the simplicity of TS FLS for developing rule based systems, researchers further extended their traditional structure with the Type-2 FL formalisms in order to accommodate higher levels of uncertainty over the system's parameters. As is discussed in [10], this transition is fairly natural since the fundamental principles of FL are independent from the nature of the membership functions. Nevertheless, little changes are required in the inference engine and defuzzifier blocks.

When developing a Type-2 TS FLS, its parameter's uncertainty can be accounted at different parts of the rule-base, namely at its antecedent ('A') or consequent ('C') level. Table 3.1 summarizes the possible combinations that can be made.

fcomputat

The literature of Type-2 TS FLSs tends to put more emphasis in the former two representations (A2-C1 and A2-C0) since they effectively make use of Type-2 FSs. However, by using Type-1 FSs at the consequent part of the rule, the A1-C1 system also accounts with higher level of uncertainty over its parameters (comparatively to the traditional Type-1 TS systems) and, for that reason, is included in the spectrum of Type-2 TS FLSs. The A2-C1 and A2-C0 TS FLSs distinguish themselves in their consequent part: in the A2-C0 case, the consequents are a linear combination of crisp values (a polynomial in its traditional sense), whereas in the A2-C1 the consequent part is a linear combination of Type-1 FSs. In the latter case, the Type-1 FS resulting from the output of each rule can be obtained by using the Extension Principle [10]. However, since the calculations necessary to obtain a crisp output value can become quite complex, their simpler interval representations are often preferred in practical applications. In the following subsections the more general A2-C1 structure will be firstly detailed, referring then to the simpler A2-C0 and A1-C1 cases.

3.3.1 A2-C1 Structure

In order to better understand how the typical FLS main blocks are implemented under a Takagi-Sugeno structure, the information processing stages of this system will be thoroughly analyzed considering IT2FSs and Interval Type-1 Fuzzy Sets(IT1FSs) at the antecedent and consequent parts of the rule base, respectively. Extending the

Table 3.1 Characterization of Type-2 TS FLSs according to the type of parameters used in the antecedent and consequent parts of the rule base

	A2–C1	A2–C0	A1–C1
Antecedent	Type-2 fuzzy sets	Type-2 fuzzy sets	Type-1 fuzzy sets
Consequent	Type-1 fuzzy sets	Crisp numbers	Type-1 fuzzy sets

generic rule described in Eq. (3.1) to the present case, an A2-C1 TS FLS rule is defined as follows:

$$\begin{aligned} R^i: \; & If\ x_1\ is\ \tilde{F}_1^i\ and\ \cdots\ and\ x_j\ is\ \tilde{F}_j^i, \\ & Then\ y^i = C_1^i x_1 + \cdots + C_j^i x_j(k)\ , \end{aligned} \tag{3.5}$$

where R^i represents the ith fuzzy rule, $\tilde{F}_j^i$ are IT2FSs, C_j^i denotes the consequent polynomial parameters given by an IT1FSs, $i = \{1, \cdots, M\}$ where M is the number of fuzzy rules, $j = \{1, \cdots, N\}$ with N representing the number of antecedents, x_j the fuzzy system inputs and y^i the rule output. Each fuzzy set C_j^i is characterized by its center (c_j^i) and spread (s_j^i) values:

$$C_j^i = [c_j^i - s_j^i; c_j^i + s_j^i]\ . \tag{3.6}$$

Despite the interval representation of the rule's consequent part parameters, ultimately the output of this stage can be summarized as two separate polynomials yielding an upper and lower bound for each rule output, represented as:

$$\begin{aligned} \overline{y}^i &= \sum_{j=1}^{N} (c_j^i * x_j + s_j^i * |x_j|) \\ , \underline{y}^i &= \sum_{j=1}^{N} (c_j^i * x_j - s_j^i * |x_j|)\ . \end{aligned} \tag{3.7}$$

The main difference between Type-2 TS FLSs and their Type-1 counterpart lies in the aggregation mechanisms used to merge the output of each rule. Similarly to the Type-2 Mamdani FLS case, to obtain the output of a Type-2 TS FLS a Type-Reduction procedure is required, in order to account with the additional degrees of freedom provided by the Type-2 FSs.

3.3.1.1 Type-Reduction

The implementation of the Type-Reduction algorithm for a IT2FLS is, in its essence, very similar to the procedure already presented in the previous chapter. Considering the more general A2-C1 TS FLS, each bound of the output interval ($\overline{y}$ and $\underline{y}$) is obtained separately by a Center-of-Sets Type-Reduction according to the upper and lower output of each rule ($\underline{y}^i$, $\overline{y}^i$) and their respective firing levels ($\underline{f}^i$, $\overline{f}^i$). To find the set of upper and lower firing levels that give the best approximation of the system's global output, each rule's output ($\underline{y}^i$ and $\overline{y}^i$) must be firstly reordered ascendantly, yielding geometric representations similar to those depicted in Fig. 3.3. This procedure is mandatory when using a Type-Reduction algorithm based on the KM principles.

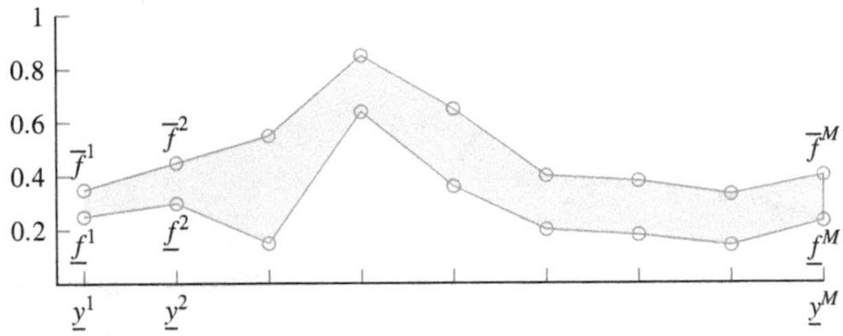

(a) Lower bound output of every system's rule ($\underline{y}^i$) sorted in ascending order.

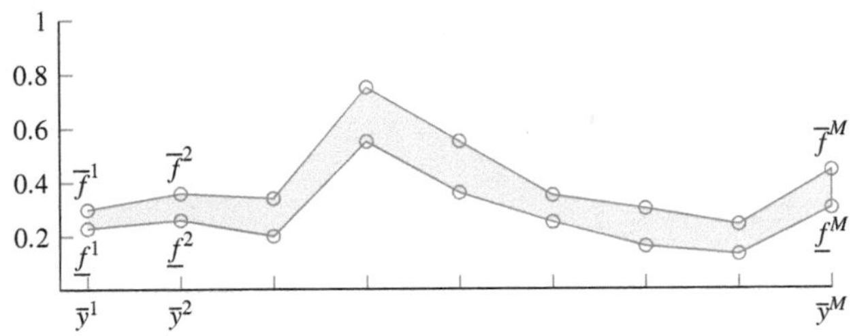

(b) Upper bound output of every system's rule ($\overline{y}^i$) sorted in ascending order.

Fig. 3.3 **a** Lower bound output of every system's rule ($\underline{y}^i$) sorted in ascending order. **b** Upper bound output of every system's rule ($\overline{y}^i$) sorted in ascending order. Polygons obtained after reordering each rule's output in order to apply the Karnik-Mendel Type-Reduction procedure

This polygon can be interpreted as a special IT2FS, as the area bounded by the rule's upper and lower firing levels in fact resemble one. Hence, the principles of Karnik-Mendel Type-Reduction can be applied to this set of points so that the optimal switching points (L and R) are found. In this procedure is usual practice to ensure that $\underline{y}^i$ and $\overline{y}^i$ have no duplicate elements, which can be easily achieved by combining the weights of duplicate elements. Then, $\underline{y}$ and $\overline{y}$ are obtained in the following equations:

$$\underline{y} = \frac{\sum_{i=1}^{L} \underline{y}^i \overline{f}^i + \sum_{i=L+1}^{M} \underline{y}^i \underline{f}^i}{\sum_{i=1}^{L} \overline{f}^i + \sum_{i=L+1}^{M} \underline{f}^i}, \tag{3.8}$$

$$\overline{y} = \frac{\sum_{i=1}^{R} \overline{y}^i \underline{f}^i + \sum_{i=R+1}^{M} \overline{y}^i \overline{f}^i}{\sum_{i=1}^{R} \underline{f}^i + \sum_{i=R+1}^{M} \overline{f}^i}, \tag{3.9}$$

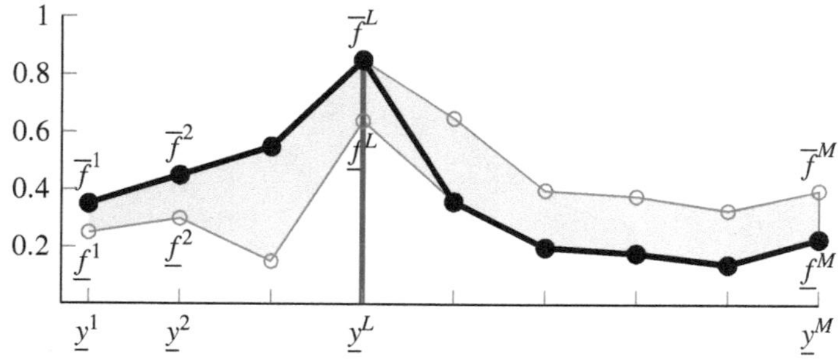

(a) Computing $\underline{y}$: Switching from the upper bounds of the firing intervals to the lower ones.

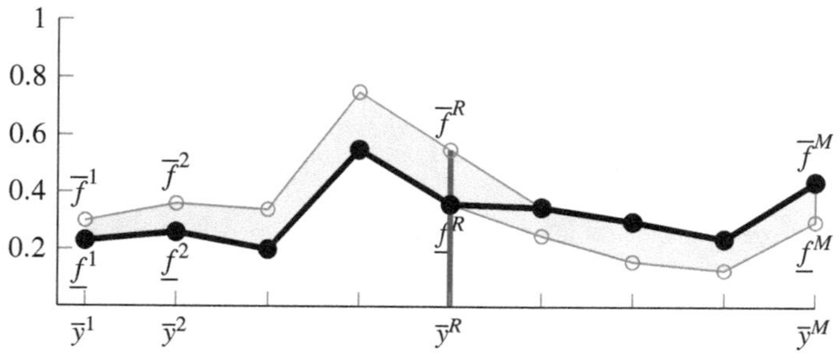

(b) Computing $\overline{y}$: Switching from the lower bounds of the firing intervals to the upper ones.

Fig. 3.4 **a** Computing $\underline{y}$: Switching from the upper bounds of the firing intervals to the lower ones. **b** Computing $\overline{y}$: Switching from the lower bounds of the firing intervals to the *upper* ones. Computation of the optimal output bound in the A2-C1 case using the Karnik-Mendel Type-reduction

where L and R are the switch points satisfying

$$\underline{y}^L \leq \underline{y} < \underline{y}^{L+1} , \tag{3.10}$$

$$\overline{y}^R \leq \overline{y} < \overline{y}^{R+1} . \tag{3.11}$$

In Fig. 3.4 the optimal Type-Reduced Fuzzy Sets are represented in bold for the upper and lower outputs:

Due to the compact representation of the Type-2 TS FLS consequents, in TS systems the Type-Reduction procedures require fewer calculations comparatively to the Mamdani case presented in the previous chapter.

3.3.1.2 Defuzzifier

After applying one of the Type-Reduction methods, the obtained Interval Fuzzy Set still has to be converted into a crisp number so that it becomes suited to the most part of the FLS application scenarios. This procedure is fairly straightforward, yielding the defuzzified value by simply computing the average of the interval's left and right endpoints:

$$y_{out} = \frac{\underline{y} + \overline{y}}{2} . \tag{3.12}$$

When such level of uncertainty representation is not necessary, the Type-2 TS FLSs can be simplified by performing some changes over the type of the antecedent or the consequent part of the FLS, as will be presented in the following sections.

3.3.2 *A2-C0 Structure*

As a particular case of the A2-C1 structure, where the consequent functions are polynomials with crisp-number parameters, the A2-C0 FLSs differ on their consequent part structure which is defined as follows:

$$\begin{aligned} R^i: \ & If\ x_1\ is\ \tilde{F}_1^i\ and\ \cdots\ and\ x_j\ is\ \tilde{F}_j^i, \\ & Then\ y^i = c_1^i x_1 + \cdots + c_j^i x_j(k) , \end{aligned} \tag{3.13}$$

where R^i represents the ith fuzzy rule, $\tilde{F}_j^i$ are IT2FSs, c_j^i denote the consequent polynomial parameters, $i = \{1, \cdots, M\}$ where M is the number of fuzzy rules, $j = \{1, \cdots, N\}$ where N is the number of antecedents, x_j are the fuzzy system inputs and y^i is the rule output.

Despite yielding just a single output for each rule, this approach also considers an uncertainty degree bounded by the rule's firing level interval. For this reason and by establishing a parallelism with the A2-C1 case, the procedure to calculate the bounds of the system output, $[\underline{y}, \overline{y}]$, is the same as presented in Eqs. (3.8) and (3.9) except that now $\underline{y}^i = \overline{y}^i = y^i$. It is important to note that, despite this equivalence, the limits R and $\overline{L}$ obtained from the Type-Reduction algorithm are not necessarily equal, as is depicted in Fig. 3.5.

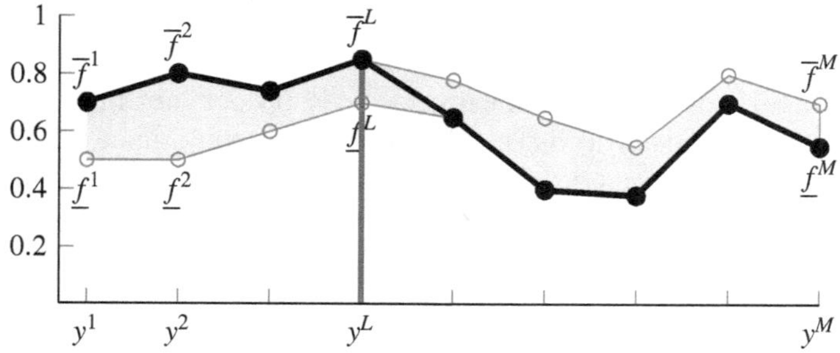

(a) Computing $\underline{y}$: Switching from the upper bounds of the firing intervals to the lower bounds.

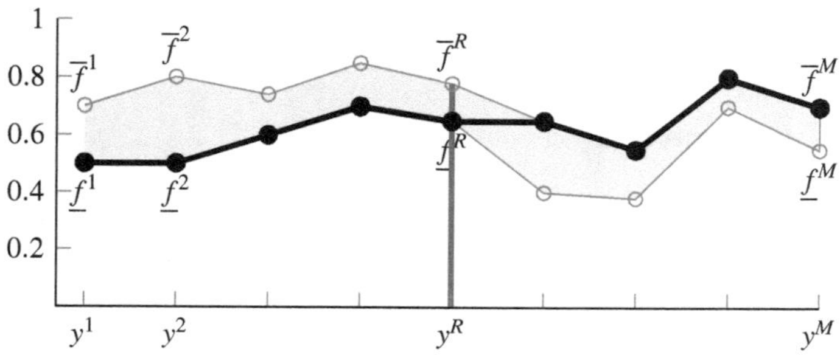

(b) Computing $\overline{y}$: Switching from the lower bounds of the firing intervals to the upper bounds.

Fig. 3.5 **a** Computing $\underline{y}$: Switching from the *upper* bounds of the firing intervals to the *lower* bounds. **b** Computing $\overline{y}$: Switching from the *lower* bounds of the firing intervals to the *upper* bounds. Computation of the optimal output bounds in the A2-C0 case using the Karnik-Mendel Type-Reduction

3.3.3 A1-C1 Structure

Considering now the last case where both antecedent and consequent part parameters' are Type-1 FSs, each FLS's rule can be written as:

$$\begin{aligned} R^i : \ & If\ x_1\ is\ F_1^i\ and\ \cdots\ and\ x_j\ is\ F_j^i, \\ & Then\ y^i = C_1^i x_1 + \cdots + C_j^i x_j(k)\ , \end{aligned} \tag{3.14}$$

where R^i represents the ith fuzzy rule, F_j^i are Type-1 FSs, C_j^i stands for the consequent polynomial parameters given by an IT1FSs, $i = \{1, \cdots, M\}$ where M is the number of fuzzy rules, $j = \{1, \cdots, N\}$ where N is the number of antecedents, x_j are the fuzzy system inputs and y^i is the rule output.

In this scenario, all the model uncertainty is treated in the consequent part of the rule. Thus, the firing level of each rule is given by a crisp number as was defined previously in Eq. (3.3). Similarly to the A2-C1 case, the output of each rule is given by an IT1FS yielding an interval bounded by $[\underline{y}^i, \overline{y}^i]$. Each values can be obtained as:

$$\overline{y}^i = \sum_{j=1}^{N} (c_j^i \, x_j + s_j^i \, |x_j|) \,,$$
$$\underline{y}^i = \sum_{j=1}^{N} (c_j^i \, x_j - s_j^i \, |x_j|) \,. \tag{3.15}$$

Since the firing levels are crisp values, the extended output of the FLS does not require the use of the Karnik-Mendel algorithm, and are simply obtained by:

$$Y = \left[\frac{\sum_{i=1}^{M} f^i \underline{y}^i}{\sum_{i=1}^{M} f^i}, \frac{\sum_{i=1}^{M} f^i \overline{y}^i}{\sum_{i=1}^{M} f^i} \right], \tag{3.16}$$

which ultimately resumes to:

$$y = \frac{\underline{y} + \overline{y}}{2} = \frac{\sum_{i=1}^{M} f^i (\sum_{j=1}^{N} c_j^i \, x_j)}{\sum_{i=1}^{M} f^i} \,. \tag{3.17}$$

Comparing the results from Eqs. (3.2) and (3.17), it is possible to conclude that the output of an A1-C1 TS FLS and the standard Type-1 TS FLS are in fact identical. For this reason, in applications where the goal is to obtain the defuzzified output of the FLS, one may choose the latter model since there is no effective advantage in implementing this more complex approach. Nonetheless, if there is interest in evaluating the uncertainty degree of the obtained output, such information can be inferred by evaluating the width of the extended output given by Eq. (3.16), which can only be derived from the A1-C1 FLS [11].

3.4 ANFIS Based on Type-2 TS Fuzzy Logic Systems

As a formal extension of the well known Type-1 TS FLS, the Type-2 TS FLSs can also be represented according to a layered architecture that best characterizes a Multi-Layer Neural System. This structure is generically depicted in Fig. 3.6 and will be presented in the sequel.

Layer 1: This layer, also known as the input layer, is defined by N nodes which embrace the crisp values relative to each input variable x_j.

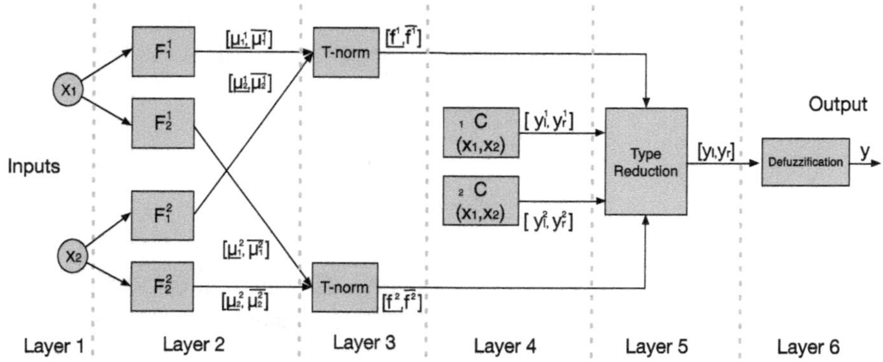

Fig. 3.6 Parallelism between the Type-2 TS FLS and the ANFIS structures

Layer 2: In this layer, the fuzzification operation is performed by evaluating the membership degree of each input variable x_j in the respective fuzzy set considered in the antecedents part of the M FLS rules. Assuming that each FS is defined by a gaussian function with fixed mean and uncertain standard deviation, as defined:

$$\tilde{F}_j^i(c_j^i, \sigma_j^i, x_j) = \exp\left[-\frac{1}{2}\left(\frac{x_j - c_j^i}{\sigma_j^i}\right)^2\right]$$
$$= G(c_j^i, \sigma_j^i, x_j),\ c_j^i \in [c1_j^i, c2_j^i]\ . \tag{3.18}$$

Unlike a Type-1 FS, where the measured membership grade is given by a number, when using IT2FSs this value is represented as an interval of uncertainty given by:

$$[\underline{\mu}_j^i, \overline{\mu}_j^i] = [G(c1_j^i, \sigma_j^i, x_j), G(c2_j^i, \sigma_j^i, x_j)] \tag{3.19}$$

Layer 3: In this layer, the upper and lower bounds of each rule firing strength are calculated. This interval is obtained by using the product *t-norm* operator [10] over the upper and lower membership values of each rule antecedents:

$$\underline{f}^i = \prod_{k=1}^{N} \underline{\mu}_k^i(x_k)\ , \tag{3.20}$$

$$\overline{f}^i = \prod_{k=1}^{N} \overline{\mu}_k^i(x_k)\ .\ . \tag{3.21}$$

At the output of this layer, it is obtained and interval $[\underline{f}^i, \overline{f}^i]$ denoting the uncertainty regarding each rule firing level.

Layer 4: Each node of the fourth layer implements the inference mechanism according to the Takagi-Sugeno principles. Considering the A2-C1 structure, the result is a linear combination of IT1FSs, yielding an interval bounded by $[\underline{y}^i, \overline{y}^i]$, whose limits are obtained by means of:

$$\overline{y}^i = \sum_{j=1}^{N} (c_j^i \; x_j + s_j^i \; |x_j|) \, , \tag{3.22}$$

$$\underline{y}^i = \sum_{j=1}^{N} (c_j^i \; x_j - s_j^i \; |x_j|) \, . \tag{3.23}$$

Layer 5: The fifth layer of the A2-C1 TS FLS is responsible for combining the output of each rule according to their upper and lower firing level bounds. This procedure is performed by using the iterative Karnik-Mendel algorithm (or one of its enhanced versions) or a closed-form approximation such as the Wu-Mendel's Uncertainty Bound Type-Reducer previously presented.

Layer 6: Finally, in the sixth layer the output of the Type-2 TS FLS is defuzzified using the average of the two endpoints $\underline{y}$ and $\overline{y}$, hence:

$$y = \frac{\underline{y} + \overline{y}}{2} \tag{3.24}$$

As was already referred, one advantage of this structure is the possibility of developing adaptation mechanisms for the model parameters based on the approximation error of the network to a input-output data dependency. The succeeding section will depict such procedures.

3.5 Training Algorithms for TS Fuzzy Systems

As was previously shown, the multi-layered architecture of TS FLSs is formally equivalent to the Feed-Forward ANN structure. Hence, the same algorithms used in ANN's training (mostly based on the output error back-propagation) are natural candidates for the development of the TS model's adaptation. In multi-layered systems, the training methods which minimize the error between the desired output and the model's output are typically implemented in two separate steps [12]. Firstly the Feed-Forward computations are performed, obtaining the values of every intermediate node of the model, followed by a backwards parameter's adaptation based on the observation of the output error. The model adaption can then be performed as a single optimization problem, by training every parameter according to the information given by the gradient and Hessian of the output error (using the Gradient descent, Gauss-Newton or Levenberg-Marquardt methods for instance [12]).

However, a more efficient and stable procedure can be alternatively employed. Since the estimation obtained by a TS system can be expressed as a weighted combination of several locally linear functions, the model training can be divided into two smaller problems. Apart from reducing the procedure's complexity for the consequent part parameters', such approach also minimizes possible numerical problems related with the larger number of estimated unknowns and the possibility of the non-linear optimization methods to be stuck into local minima. The referred approach, known as Hybrid Training [2] is performed as follows:

- At a given sampling instant, considering the parameters of the model's antecedent part fixed, the output of the TS Fuzzy model results from the weighted contribution of several linear models according to the firing level of their respective rules. Therefore, the consequent part parameters can be trained using a least squares method such as the Recursive Least Squares (RLS).
- Afterwards, by fixing the consequent parameters, the non-linear part of the model can be trained by back-propagating the output error to each one of the antecedent parameters using methods based on the error signal derivatives.

As was clear from the previous sections, the use of Type-2 FL concepts over the TS FLSs yield a significant amount of additional unknown parameters that must be estimated. Their optimal values can be obtained either by directly employing optimization algorithms [13, 14], or using recursive trained algorithms [15, 16]. Yet, considering a Type-2 TS Fuzzy model estimation solely as an error minimization problem misses the initial purpose of embedding uncertainty intervals over Type-1 FLS and, when applied without supervision, may result in FOU that no longer have a valid meaning for the model's interpretability. Hence, as every membership function is ultimately obtained by varying one or several parameters of a Type-1 FS, the training methods further presented focus on the Type-1 TS model structure. The obtained parameters can then be used as a starting point for the development of its Type-2 TS Fuzzy model, by expanding the uncertainty intervals by a factor so that the overall model performance is improved. Despite the simplicity and intuitiveness of this approach, it is currently not discussed in literature. Nonetheless, it is fairly simpler than the Type-2 TS training procedures currently available and it was found to provide superior numerical robustness in the development of model based controllers [17].

In the following sections, the local training procedures used in the development of a Type-1 TS FLS will be presented.

3.5.1 Model Initialization

The initialization of the antecedent part of the Type-1 FLS plays an important role in the definition of the system's structure as it will ultimately set the minimum number of rules necessary to approximate the input-output dependencies of the system. Since its appropriate dimension is hardly known at the beginning of the design stage, it is common practice to use one-pass clustering algorithms over a large input-output dataset in order to extract natural groupings of data from it. Clustering is usually employed in classification problems and it is also often adopted as an initialization procedure of the FLSs' rule base. Despite the differences in nomenclature and information organization, a n-dimensional cluster is functionally equivalent to the antecedent part of a rule with n input variables. Therefore, such approach can be used to obtain the appropriate number of rules as well as defining the center and variance of each membership function of the model's input space.

Literature is rich in clustering algorithms that, ultimately, represent variations of the original Fuzzy C-Means (FCM) clustering algorithm [18]. The FCM algorithm is an iterative optimization method used to find the optimal centers of the membership functions that partition the input space of a FLS, aiming to minimize the cost function:

$$J = \sum_{k=1}^{K} \sum_{i=1}^{C} \mu_{ik}^{m} \|x_k - v_i\|^2 , \tag{3.25}$$

where K is the number of data points, C is the number of clusters, x_k is the kth n-dimensional data point, v_i is the ith cluster center, μ_{ik} is the degree of the membership of the kth data in the ith cluster and m is a constant greater than 1 (typically $m = 2$) that defines the width of the cluster. Provided the desired number of clusters and an initial guess for each cluster center v_i, the FCM algorithm will converge to a solution which represents either a local or global minimum of the given cost function.

As in every non-linear optimization problem, the quality of the solution found is highly related with the choice of the initial values of the clusters' centers. Using the Mountain Method [19], such constraint is overcome by simply using a grid partition of a n-dimensional input space as a starting point for the clusters' parameters. However, the computational complexity of such approach can escalate very easily, growing exponentially with the number of input variables of the system. The Subtractive Clustering algorithm [20] circumvents the dimensionality issues of the previous method by considering each data point (and not each one of the possible grid partitions) as a potential cluster center. The computational complexity depends on the dimensionality of the data set but, more importantly, is unrelated with the input space dimensionality.

In the Subtractive Clustering algorithm, the possible cluster centers are found according to a metric that evaluates the potential of each data point in assuming such role. Such metric is presented in the equation:

$$P_i = \sum_{j=1, j \neq i}^{n} \exp\left[-\alpha \|x_i - x_j\|^2\right] , \tag{3.26}$$

where

$$\alpha = \frac{4}{r_a^2} \tag{3.27}$$

and r_a is a positive constant related with the radius of influence of each possible cluster center candidate. Thus, the points with higher number of neighbor points will present an higher potential value, having more chances to be selected as cluster centers. After calculating the potential of every point, the data point with higher potential value is selected as the first cluster. Let x_1^* be the location of the first cluster center and P_1^* its potential. The potential of every remaining data point is revised by the equation:

$$P_i \Leftarrow P_i - P_1^* \exp\left[-\beta\|x_i - x_1^*\|^2\right] , \tag{3.28}$$

where

$$\beta = \frac{4}{r_b^2} \tag{3.29}$$

and r_b is a positive constant. This second step effectively penalizes the data points closer to the first cluster reducing their potential of be selected as cluster centers in the successive iterations of the algorithm. To avoid the selection of closely spaced clusters, r_b should be greater than r_a usually in the proportion of $r_b = 1.5r_a$. This procedure is repeated until the potential of the kth cluster is a small fraction of the first cluster extracted, as follows:

$$P_k^* < \varepsilon P_1^* . \tag{3.30}$$

The value of this threshold, ε, will define the number of data points accepted as cluster centers and set the dimensionality of the rule base.

As far as it concerns the variance of the gaussian fuzzy MF, its value can be obtained considering the equivalence between Eq. (3.31) and the clustering metric presented in Eq. (3.26). Thus, a gaussian membership function can be defined as:

$$F(x, c, \sigma) = \exp\left[-\frac{(x-c)^2}{2\sigma^2}\right], \tag{3.31}$$

and the variance of the membership functions obtained via Subtractive Clustering is given by:

$$\sigma^2 = \frac{r_a^2}{8} . \tag{3.32}$$

While the Subtractive Clustering algorithm significantly contributed to the development of smaller yet well performing TS systems, not every input variable is strictly relevant for an accurate non-linear input space partitioning. If all input variables present at the consequent part regressive model are considered, one can easily end up in a combinatorial problem which leads to a very large structure and even reduce its extrapolation capabilities. Therefore, it is of great importance to establish a balance between the model accuracy and its complexity. Based on this premise [21] proposed a methodology where only the regressors having a non-linear impact in the parameters of the consequent part of the TS systems are considered in the rule extraction procedure. In fact, those are the state variables that define the necessity of employing different linear approximations over different operating scenarios. Opposing to the Mamdani inference procedure, pruning the antecedent part of a TS structure does not necessarily impairs its approximation capability since the consequent one inherently establishes the output interdependency with every input variable.

3.5.2 *Training of the Antecedent Part of the Rule Base*

After the model initialization stage, the parameters of the antecedent part can be fine-tuned using a non-linear optimization algorithm such as the Gradient Descent or the Levenberg-Marquardt [22, 23]. Since the input space of a TS Fuzzy Model is less likely to present significant variations over the time, the initial antecedent parameters estimations are usually fairly close to the optimal ones. Thus, while it may be argued that the Gradient Descent training might present a slower convergence towards the optimal solution than Hessian based methods such as the LM, for the majority of applications this is not a restrictive drawback as the adaptiveness of the model resides mostly at its consequent part. Therefore, the Gradient Descent update rules for the variance and the center of each antecedent part of Type-1 FS can be obtained based on the minimization of the squared error of the prediction model as:

$$c^i_j(k+1) = c^i_j(k) - \eta \frac{\partial E}{\partial c^i_j}\,, \tag{3.33a}$$

$$\sigma^i_j(k+1) = \sigma^i_j(k) - \eta \frac{\partial E}{\partial \sigma^i_j}\,, \tag{3.33b}$$

where E is the prediction error and η is the learning coefficient, usually chosen in the interval $0 < \eta \leq 0.2$ [12].

The partial derivatives of each free parameter present in the antecedent part of the rule base are obtained as considering

$$E = \frac{1}{2}\sum_{k=1}^{K}(y^d(k) - y(k))^2 \tag{3.34}$$

and

$$G(c^i_j, \sigma^i_j, x_j) = exp\left[-\frac{1}{2}\left(\frac{x_j - c^i_j}{\sigma^i_j}\right)^2\right]. \tag{3.35}$$

Therefore,

$$\frac{\partial E}{\partial \sigma_j^i} = \frac{\partial E}{\partial y} \frac{\partial y}{\partial f^i} \frac{\partial f^i}{\partial \mu_j^i} \frac{\partial \mu_j^i}{\partial \sigma_j^i} , \tag{3.36}$$

$$\frac{\partial E}{\partial c_j^i} = \frac{\partial E}{\partial y} \frac{\partial y}{\partial f^i} \frac{\partial f^i}{\partial \mu_j^i} \frac{\partial \mu_j^i}{\partial c_j^i} , \tag{3.37}$$

where

$$\frac{\partial E}{\partial y} = -1 , \tag{3.38}$$

$$\frac{\partial y}{\partial f^i} = \frac{y^i \sum_{i=1}^{M} f^i - \sum_{i=1}^{M} f^i y^i}{\left(\sum_{i=1}^{M} f^i \right)^2} , \tag{3.39}$$

$$\frac{\partial f^i}{\partial \mu_j^i} = \prod_{k=1, k \neq j}^{N} \mu_k^i , \tag{3.40}$$

$$\frac{\partial \mu_j^i}{\partial c_j^i} = \frac{\left(x_j - c_j^i \right)}{(\sigma_j^i)^2} G(c_j^i, \sigma_j^i, x_j) , \tag{3.41}$$

and

$$\frac{\partial \mu_j^i}{\partial \sigma_j^i} = \frac{\left(x_j - c_j^i \right)^2}{(\sigma_j^i)^3} G(c_j^i, \sigma_j^i, x_j) . \tag{3.42}$$

At this stage, the importance of the Forward Pass for the training of every antecedent part parameter is clearer, since the values of parameters μ_j^i and f^i depend on the execution of one iteration of the TS FLS for a given a set of input values.

3.5.3 *Training of the Consequent Part of the Rule Base*

When a system to be identified is linear on its parameters, procedures based on the squared error minimization such as the RLS algorithm are know to provide the best convergence to the solution which better approximates a specific input/output behavior [24]. The TS FLSs fall into this category since, by considering that at given instant the antecedent part firing levels' are constant, the output of the system is no more than a weighted combination of linear functions given by each rule's consequent part.

The training procedure for the consequent part of the rule base can be translated into a least squares numerical optimization problem according to two different ways, using either a global or a local optimization approach. The global approach leads to:

$$\theta = arg\ min \sum_{k=1}^{K} \left(y_k^* - \sum_{n=1}^{N} f^i \varphi \hat{\theta}_i \right) , \tag{3.43}$$

where y^* denotes the system output to approximate, φ_i is the n-dimensional observation vector, $\theta = [\theta_1\ \theta_2\ \cdots\ \theta_N]$ represents the concatenation of all the individual rule's parameter vectors, f_i is the normalized firing level of each rule, N is the number of rules of the system and K is the length of the training dataset.

Considering now the local training approach, the model parameters are obtained by:

$$\theta_i = arg\ min \sum_{k=1}^{K} \left(f^i y_k^* - f^i \varphi \hat{\theta}_i \right) , \tag{3.44}$$

where y^* is the system output to approximate, φ is the n-dimensional observation vector, and θ_i is the parameter vector of each individual rule, f^i is the normalized firing level of each rule, N is the number of rules of the system and K is the length of the training dataset.

In the former approach, employed for example in [15], every consequent functions parameters' are trained as a whole in a single regression problem, while in the latter case the training of the consequent part of each rule constitutes a separate optimization problem. In terms of error minimization, the method used is not crucial but, if each rule output is to be interpreted as a local model, then the employed approach ultimately defines its usability. As is argued in [25, 26], a globally optimal model by no means guarantees a locally adequate behavior of the sub-models that constitute the TS structure, often leading to over-fitting problems and meaningless parameters estimates which can ultimately result in numerical instability as verified in [15]. A simple interpretation of this problem is depicted in Fig. 3.7.

From the depicted scenario, the model estimated by local optimization properly describes the local behavior of the function, despite giving a less accurate global fitting. For the global optimization approach, the opposite holds—a better global fit is obtained but the consequent part functions are not relevant for a local description of the system's behavior. As will be clearer in the further chapters, the validity of the local models will be important to perform the synthesis of model based controllers. To comply with this requirement, constrained and multi-criteria optimization methods can be applied over the global training approach [26] in order to restrict the parameters domain of freedom. However, the training procedure becomes a quadratic optimization problem instead of a least squares one, increasing both the complexity and the

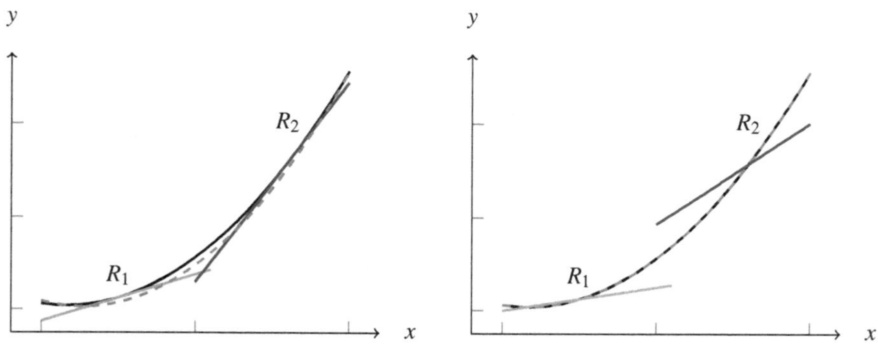

Fig. 3.7 The result of local (*left*) and global (*right*) optimization of the consequent parameters for a single input two rule's system. The dashed line is the system output

required computational effort. For this reason, establishing a compromise between modeling accuracy and training method complexity, the local training approach will be followed in the present book using a weighted RLS algorithm.

While the RLS algorithm provides an efficient method of performing the local training of each rule consequent part, it is known that its conventional formulation lacks the required adaptability to track time varying parameters [24] since it gives the same importance to all the previous samples for the current time estimation. To overcome this issue, modified versions of the cited algorithm introduced an exponential weight that reduces the significance of past samples according to their obsolescence [24]. Yet, in practical scenarios, where the system excitation is insufficient or not uniform over the whole parameters' space, this information loss mechanism can lead to numerical stability problems due to a phenomenon referred in the literature as covariance matrix windup [24]. Therefore, several heuristics have been proposed to overcome this problem either by adjusting the algorithm forgetting factor considering the evolution of the estimation error or by monitoring the evolution of the covariance matrix [24]. Approaches based on the latter method are considered more robust and, among the existing methods, the Directional Forgetting mechanism [27, 28] stands out for its simplicity, stability and capability of maintaining the adaptability of the estimator to fast and slow parameter's variations. Despite its superiority over long training epochs, Hybrid TS model learning techniques continue to put emphasis in methods that periodically reset the covariance matrix of the estimator to maintain is stability and adaptability. Therefore, it represents an approach that can be simply interpreted as a "reset" of the estimator. Only few publications tackle this problem using the Recursive Least Squares with Directional Forgetting (RLSDF) algorithm [29, 30] but, given its proven robustness, it will be used in the examples presented in this book and will be following defined.

Considering that, at each sampling instant, the estimated output results from the weighted contribution of several linear models according to each rule's normalized firing level, f^i, the output of the model is given by:

$$y = \frac{\sum_{i=1}^{M} f^i y^i}{\sum_{i=1}^{M} f^i} \,. \tag{3.45}$$

Following a local training based approach, the cost function J to be minimized is defined as the sum of the cost functions J_i of each sub-model, as:

$$J = \sum_{i=1}^{R} J_i \,, \tag{3.46}$$

where

$$J_i = (y^* - \varphi^T \hat{\theta}_i) f^i \lambda_i (y^* - \varphi^T \hat{\theta}_i) \,. \tag{3.47}$$

Expression 3.47 can ultimately be rewritten as:

$$J_i = (\sqrt{f^i} y^* - \sqrt{f^i} \varphi^T \hat{\theta}_i) \lambda_i (\sqrt{f^i} y^* - \sqrt{f^i} \varphi^T \hat{\theta}_i) \,, \tag{3.48}$$

where, for the ith system's rule, f^i is the normalized activation level, φ the n-dimensional observation vector, $\hat{\theta}_i$ the n-dimensional parameter's vector of the locally linear sub-model and λ_i a weight related to the estimator forgetting factor. Assuming that the model has R rules, the same number of linear models must be estimated and, consequently, the same number of cost functions must be optimized.

At the time instant k, the parameters of the discrete-time linear predictor for the ith rule can be recursively obtained using the weighted RLSDF algorithm as follows:

$$\hat{\theta}_i(k) = \hat{\theta}_i(k-1) + K(k)\varepsilon_i(k) \,, \tag{3.49a}$$

$$\varepsilon_i(k) = \sqrt{f^i} y^*(k) - \sqrt{f^i} \varphi(k)^T \hat{\theta}_i(k) \,, \tag{3.49b}$$

$$r_i(k) = f i \varphi^T(k) P_i(k-1) \varphi(k) \,, \tag{3.49c}$$

$$K(k) = \frac{\sqrt{f^i} P_i(k-1)\varphi(k)}{1 + r_i(k)\alpha_i(k)} \,, \tag{3.49d}$$

$$P_i(k) = P_i(k-1) - K(k)\sqrt{f^i} \varphi^T(k) \alpha_i(k) P_i(k-1) \,, \tag{3.49e}$$

$$\alpha_i(k) = \begin{cases} \lambda_i - \dfrac{1-\lambda_i}{r_i(k)}, & r_i(k) > 0 \\ 1, & r_i(k) = 0 \end{cases}. \tag{3.49f}$$

The λ_i parameter represents the algorithm forgetting factor and it is usually chosen in the interval $0.95 \leq \lambda_i \leq 1$ [27]. The value of λ_i establishes a commitment regarding the algorithm's capability in tracking fast/slow variations of the model parameters.

3.6 Conclusions

The development of rule-base systems according to the Takagi-Sugeno structure significantly expanded the domains of applicatbility of Fuzzy Logic Systems due to their closeness to well known linear modeling theory and the use of simple training algorithms (derived for their equivalent ANFIS structure). By inheritance of Type-1 Takagi-Sugeno Fuzzy Logic Systems' main properties and due to their information uncertainty representation features, Type-2 Fuzzy Logic Systems based on the Takagi-Sugeno structure have a superior potential to excel in system modeling tasks.

Since every Type-2 Takagi-Sugeno Fuzzy Logic Systems is ultimately defined by embedding a Footprint-of-Uncertainty over its Type-1 counterpart parameters, the procedure of defining a Type-2 Fuzzy Logic System is significantly simplified if one focus on the centers of the uncertainty intervals. Such approach seems to be disregarded by the related literature but, despite its simplicity, it is not less valid than the currently available ones. In fact, it provides to the practitioner a deeper insight regarding the influence of the uncertainty factors on the quality and accuracy of the developed systems and is computationally less demanding due to the smaller number of parameters that must be tuned. Such dependency will be clearer in the succeeding chapters, by studying the applicability of Type-2 Fuzzy Logic Systems in modeling and control applications.

References

1. Takagi, T., Sugeno, M.: Fuzzy identification of systems and its applications to modeling and control. IEEE Trans. Syst. Man Cybern. **15**(1), 116–132 (1985)
2. Jang, J., Sun, C., Mizutani, E.: Neuro-Fuzzy and Soft Computing–A Computational Approach to Learning and Machine Intelligence. Prentice-Hall (1999)
3. Lin, T., Lee, C.: Neural network based fuzzy logic control and decision system. IEEE Trans. Comput. **40**(12), 1320–1336 (1991)
4. Berenji, H., Khedkar, P.: Learning and tuning fuzzy logic controllers through reinforcements. IEEE Trans. Neural Netw. **1992**(3), 724–740 (1992)
5. Nauck, D., Kurse, R.: Neuro-fuzzy systems for function approximation. In: 4th International Workshop Fuzzy-Neuro Systems (1997)

6. Juang, F., Chin Lin, T.: An on-line self constructing neural fuzzy inference network and its applications. IEEE Trans. Fuzzy Syst. **1998**(6), 12–32 (1998)
7. Jang, R.: Neuro-fuzzy modeling: architecture, analyses and applications, Ph.D. Thesis, Department of Electrical Engineering and Computer Science, University of California, Berkley, (1992)
8. Abraham, A., Nath, B.: Hybrid Intelligent Systems: A Review of a decade of Research. School of Computing and Information Technology, Faculty of Information Technology, Monash University, Australia, Technical Report Series 5(2000), pp. 1–55 (2000)
9. Benzaouia, A., El Hajjaji, A.: Advanced Takagi–Sugeno Fuzzy Systems, Studies in Systems, Decision and Control, vol. 8. Springer (2014)
10. Mendel, J.: Uncertain Rule-Based Fuzzy Logic Systems: Introduction and New Directions. Prentice-Hall (2001)
11. Boumella, N.; Djouani, K.; Boulemden, M.: On an interval type-2 TSK FLS A1-C1 consequent parameters tuning. In: IEEE Symposium on Advances in Type-2 Fuzzy Logic Systems, pp. 150–156 (2011)
12. Nørgaard, M., Ravn, O., Poulsen, K., Hansen, L.: Neural Networks for Modeling and Control of Dynamic Systems. Springler, London (2000)
13. Kumbasar, T.; Eksin, I.; Guzelkaya, M.; Yesil, E.: Type-2 fuzzy model inverse controller design based on BB-BC optimization method. In: 18th World Congress of the International Federation of Automatic Control (IFAC) (2011)
14. Martínez-Sotoa, R., Castillo, O., Aguilar, L.: Type-1 and Type-2 fuzzy logic controller design using a hybrid PSO–GA optimization method. In: Information Sciences – Processing and Mining Complex Data Streams, vol. 285, pp. 35–49. Elsevier (2014)
15. Chia-Feng, J., Yu-Wei, T.: A self-evolving interval Type-2 fuzzy neural network with online structure and parameter learning. IEEE Trans. Fuzzy Syst. **16**(6), 1411–1423 (2008)
16. Interval singleton type-2 TSK fuzzy logic systems using orthogonal least-squares and back-propagation methods as hybrid learning mechanism. In: 2011 11th International Conference on Hybrid Intelligent Systems, pp. 417–423 (2011)
17. Antão, R., Mota, A., Martins, R.: Model-based Control using Interval Type-2 Fuzzy Logic Systems. Springer, Soft Computing (2016)
18. Bezdek, J., Hathaway, R., Sabin, M., Tucker, W.: Convergence theory for fuzzy c-means: counterexamples and repairs. IEEE Trans. Syst, Man Cyber. **17**(5), 873–877 (1987)
19. Yager, R., Filev, D.: Approximate clustering via the mountain method. IEEE Trans. Syst. Man Cybern. **24**(8), 1279–1284 (1994)
20. Chiu, S.: Fuzzy model identification based on cluster estimation. J. Intell. Fuzzy Syst. **2**(3) (1994)
21. Su, M., Rhinehart, R.: A generalized TSK model with a novel rule antecedent structure: structure identification and parameter estimation. Comput. Chem. Eng. **34**(8), 1199–1219 (2010)
22. Levenberg, K.: A method for the solution of certain problems in least squares. Q. Appl. Math. **2**, 164–168 (1944)
23. Marquardt, D.: An algorithm for least-squares estimation of non-linear parameters. J. Soc. Ind. Appl. Math. **11**(2), 431–441 (1963)
24. Ljung, L.: System Identification: Theory for the User. Prentice Hall (1987)
25. Yen, J.; Wang, L., Gillespie, C.: Improving the interpretability of TSK fuzzy models by combining global learning and local learning. IEEE Trans. Fuzzy Syst. **6**(4) (1998)
26. Babuška, R., Verbruggen, H.: Neuro-fuzzy methods for nonlinear system identification. Ann. Rev. Control, **27**, 73–85. Pergamon (2003)
27. Hägglund, T.: New estimation techniques for adaptive control, Ph.D. Thesis, Department of Automatic Control, Lund Institute of Technology, Sweden (1983)
28. Kulhavý, R.: Restricted exponential forgetting in real-time identification. In: IFAC Symposium on Identification and System Parameter Estimation, pp. 1143-1148 (1985)
29. Mendes, J., Araujo, R., Souza, F.: Adaptive fuzzy identification and predictive control for industrial processes. Expert Syst. Appl. **40**, 6964–6975. Elsevier (2013)
30. Bouillon, M., Anquetil, E., Almaksour, A.: Decremental learning of evolving fuzzy inference systems: application to handwritten gesture recognition. In: Machine Learning and Data Mining in Pattern Recognition, Lecture Notes in Computer Science, vol. 7988, pp. 115–129 (2013)

Chapter 4
System Modeling Using Type-2 Takagi-Sugeno Fuzzy Systems

4.1 Introduction

The development of computational models capable of accurately describe a process's dynamic response is a task ultimately dependent on its physical phenomena complexity and the type of disturbances that may affect its operation. According to the literature [1], there exist several different approaches to obtain such system's description:

- Physical models
- Black-Box models
- Grey-Box models

When a deep knowledge about the physical laws underlying the system's behavior is present, it becomes possible to develop models based on a set of differential equations describing the rate of change of the system's state variables. The development of fundamental models typically requires a large number of parameters but, once each one of them is available, either estimated from experimental scenarios or simply by being well known physical constants, it becomes possible to extrapolate results from a large range of operation regions. However, many times such procedure become a time demanding task, requiring specific test conditions so the influence of individual parameters is isolated. This condition is even more problematic when the system presents non-linear behavior since the superposition principle is no longer valid. Consequently, a large number of possible combinations of input variables has to be necessarily tested, ultimately posing a large search space for numerical optimization algorithms to obtain the set of parameter that best approximate the process response. For this reason, the use of physical modeling methods in the development of prediction models and controller synthesis is very restricted.

In what concerns the parameters' interpretability, Black-Box models advocate the opposite paradigm in a model development procedure in the sense that the it is developed on an information processing perspective, without any considerations regarding the process's physical properties. As so, a mathematical description of

R. Antão, *Type-2 Fuzzy Logic*, Nonlinear Physical Science,
DOI 10.1007/978-981-10-4633-9_4

the system is developed aiming to find relationships among variables that best fits the input/output data obtained from the process. Typically, such descriptions can be represented using polynomial structures, either based on linear Auto-Regressive Moving Average with eXogenous input (ARX) and Auto-Regressive with eXogenous inputs (ARX) structures, or on Non-Linear Auto-Regressive with eXogenous input (NARX) such as Feed-Forward Neural Networks, Wiener or Hammerstein models [1]. In any of its forms, an input-output model can be generically represented as:

$$\begin{aligned}\hat{y}(k) = f(y(k-1), \cdots, y(k-n_y), u(k-d-1), \cdots, \\ u(k-d-n_u), v(k-1), \cdots, v(k-n_c))\,,\end{aligned} \tag{4.1}$$

where f defines the variables' interdependency, $\hat{y}$ is the predicted output, y are the process's past outputs, and its exogenous inputs are given by the actuation values u and the measured external disturbances v. The parameters n_y, n_u, n_c are chosen according to the relevancy of the past values to the present estimation and d represents the process's dead-time.

Obtaining a Black-Box model is a procedure which encompasses several stages, typically grouped as: structure selection, parameter identification and model validation [2]. Due to its iterative nature, such procedure is not a single-pass one, and typically requires several experiments before a robust yet simple model is obtained.

There may exist scenarios where the process's dynamic behavior is totally unknown and a Black-Box modeling approach is ultimately used to develop a model. Nevertheless, most of the times the little knowledge available about the process can significantly reduce the number of iterations required to find an optimal model. By combining the fundamental and Black-Box modeling principles, an approach known as Grey-Box modeling emerge. Empirical relations (which usually have a limited region of validity) can be established among the relevant system variables, reducing the model structure size while closely approximate the plant dominant dynamics in a specific operation region. Knowing how disturbances affect its operation, the noise's spectral distribution, the non-linearities over the expected operation regime or the memory that the system has about past inputs or outputs are some of the parameters that may give important hints to accomplish this task [2].

The parameters of a Grey-Box model do not present any physical meaning. Nevertheless, when such models are developed according to linear systems' identification theory, important relations can be established between their parameters and the dynamic response of the system (as stability and transient response analysis) [3]. For this reason, despite the improved modeling capabilities of non-linear approaches, linear strategies continue to have a large adoption due to their simpler structure and easier theoretical analysis. As was presented in Chap. 3, Takagi-Sugeno FLSs stand in between both methods, providing a simple framework to approximate non-linear input-output relations based on the use of several locally linear dynamic models. Considering their use in process modeling applications, the linear consequent part of the model can also be developed according to a Grey-Box principles, thus extending important developments from linear modeling and control theory to TS FLSs.

4.2 Locally Linear Models Based on Type-2 TS Fuzzy Logic Systems

The use of TS FLSs in system modeling applications is a natural process since a dynamic interpretation of its structure is obtained by replacing its input variables with the relevant regression variables. Considering the particular case of developing linear systems based on the ARX structure, the consequent part of the TS model is commonly represented as:

$$A(z^{-1})y(k) = B(z^{-1})z^{-d}u(k) + C(z^{-1})v(k) , \tag{4.2}$$

where $u(k)$ and $y(k)$ are the control and output sequences of the plant, d is the dead time of the system and $v(k)$ is white noise. The symbols A, B, and C represent the polynomials represented in the backward shift operator z^{-1}:

$$\begin{aligned} A(z^{-1}) &= 1 + a_1 z^{-1} + a_2 z^{-2} + \cdots + a_{n_a} z^{-n_a} , \\ B(z^{-1}) &= b_1 z^{-1} + b_2 z^{-2} + \cdots + b_{n_b} z^{-n_b} , \\ C(z^{-1}) &= 1 + c_1 z^{-1} + c_2 z^{-2} + \cdots + c_{n_c} z^{-n_c} , \end{aligned} \tag{4.3}$$

where the parameters n_a and n_b are related to the order of the estimated ARX model and n_c related with the expected type of disturbance.

As was already referred, the interpolation features provided by TS descriptions provide good approximations of non-linear behaviors as long an adequate number of rules are employed. Nonetheless, in modeling applications, the parameters of each locally linear approximations used at every rule's consequent part will be very likely affected by uncertainty factors due to time varying conditions of the modeled processes. Type-2 TS FLSs inherently encode in their structure the mechanisms necessary to represent such variability and their inclusion in system modeling applications is performed by assuming the existence of a uncertainty factor over the parameters of a Type-1 TS Fuzzy model. The antecedent part of the model can be extended by varying either the variance or the mean value of the membership functions that partition the model's input space. The consequent parameters can be generically defined in an interval representation $\tilde{p}$, as:

$$\tilde{p} = [p - s; p + s] \equiv [\underline{p}; \overline{p}] , \tag{4.4}$$

where the center of the interval (p) is given by the coefficients of the polynomials A, B and C presented in (4.3), and s is the width of the considered uncertainty interval, which can be defined as a percentage of the parameter p. Ultimately, a Type-2 TS Fuzzy Model can represent each rule consequent by two polynomial functions associated with the upper and lower bounds of its parameters. Together with the Type-Reduction mechanisms, such representation constitute a compact and effective way of embedding several possible models that best approximate a specific operation point of a process. A generic definition of the ith rule of a system with M *If-Then* rules and N antecedents is presented as follows:

$$\begin{aligned} If \quad & y(k-1)\ is\ \tilde{F}_1^i\ and\ \cdots\ and\ u(k-1)\ is\ \tilde{F}_N^i, \\ Then \quad & \begin{array}{ll} \underline{A}^i(z^{-1})\underline{y}^i(k) & = \underline{B}_i(z^{-1})u(k) + \underline{C}_i(z^{-1})v(k) \\ \overline{A}^i(z^{-1})\overline{y}^i(k) & = \overline{B}_i(z^{-1})u(k) + \overline{C}_i(z^{-1})v(k) \end{array}, \end{aligned} \tag{4.5}$$

and:

$$[\underline{A}^i, \overline{A}^i](z^{-1}) = [1, 1] + [\underline{a}_{1i}, \overline{a}_{1i}] + [\underline{a}_{2i}, \overline{a}_{2i}]z^{-1} + \cdots + [\underline{a}_{n_a i}, \overline{a}_{n_a i}]z^{n_a}\ , \tag{4.6a}$$

$$[\underline{B}^i, \overline{B}_i](z^{-1}) = [\underline{b}_{1i}, \overline{b}_{1i}]z^{-1} + \cdots + [\underline{b}_{n_b i}, \overline{b}_{n_b i}]z^{n_b}\ , \tag{4.6b}$$

$$[\underline{C}^i, \overline{C}_i](z^{-1}) = [1, 1] + [\underline{c}_{1i}, \overline{c}_{1i}]z^{-1} + \cdots + [\underline{c}_{n_c i}, \overline{c}_{n_c i}]z^{n_c}\ . \tag{4.6c}$$

4.2.1 Development of the Interpolated Interval Type-2 Fuzzy Model

One of the advantages upcoming from the existence of a system's model is the possibility of developing control techniques capable of adapting their performance to different operating conditions. Model based control techniques such as Pole-Placement [4] or Model Predictive Control [5] are some examples of such approach. From the point of view of the system's behavior prediction, TS FLSs provide a simple mechanism of interpolating its output based on the weighted contribution of several locally linear models. However, when a model is used in a model-based control framework, it is important not only to have the predicted output but, also the parameters of an equivalent model. Fortunately, since the TS FLSs' rule aggregation is of linear nature, one can obtain a single linear model where each parameter results from the weighted contribution of every locally linear model partially activated at the current operating region [6]. Naturally, transitory regions that activate several partitions of the input space result from the contribution of several models and, therefore, present higher levels of uncertainty. For this reason, in [7] this issue is overcame and the model's accuracy is improved by merging a set of the best performing linear models at each region obtained by a genetic algorithm. By combining multiple TS fuzzy models, one can improve the overall identification performance of the model, since the expected error upcoming from the use of multiple models will (in the worst case scenario) not exceed the expected error of the individual models [8]. The present work aims to extend such principle by using the model uncertainty representation according to the Type-2 TS FLSs' principles.

Since a Type-2 FLS output is ultimately represented by a bounded output, two average models of the plant ($\tilde{\overline{y}}(k)$ and $\tilde{\underline{y}}(k)$) will be obtained. Hence, defining the upper and lower bounds separately and having the C polynomial equal to unity (leading to an ARX model) the average estimated model ($\tilde{y}$) is represented as:

$$\tilde{y}(k) = [1 - \tilde{A}(z^{-1})]y(k-1) + \tilde{B}(z^{-1})u(k) \tag{4.7a}$$

where,

$$\tilde{A}(z^{-1}) = 1 + \tilde{a}_1 z^{-1} + \tilde{a}_2 z^{-2} + \cdots + \tilde{a}_{n_a} z^{n_a} , \tag{4.8a}$$

$$\tilde{B}(z^{-1}) = \tilde{b}_1 z^{-1} + \cdots + \tilde{b}_{n_b} z^{n_b} . \tag{4.8b}$$

The coefficients of the polynomials $\tilde{A}(z^{-1})$ and $\tilde{B}(z^{-1})$ are obtained by averaging the M consequent models, separately for the upper and lower bounds, according to their normalized firing level. This procedure is presented in the equations:

$$\tilde{a}_k = \frac{\sum_{i=1}^{M} w^i a_k^i}{\sum_{i=1}^{M} w^i}, \qquad k = 1, \cdots, n_a , \tag{4.9a}$$

$$\tilde{b}_k = \frac{\sum_{i=1}^{M} w^i b_k^i}{\sum_{i=1}^{M} w^i} \qquad k = 0, \cdots, n_b . \tag{4.9b}$$

For the lower average model, the weights w that define the contribution of each sub-model for the expression of $\tilde{y}(k)$ and the parameters a_k^i and b_k^i are given by:

$$w \equiv \underline{w} = [\overline{f}^1, \overline{f}^2, \cdots, \overline{f}^L, \underline{f}^{L+1}, \cdots, \underline{f}^M], \ a_{ki} \equiv \underline{a}_{ki}, \ b_{ki} \equiv \underline{b}_{ki} , \tag{4.10}$$

while for the upper average model, the parameters are given by:

$$w \equiv \overline{w} = [\underline{f}^1, \underline{f}^2, \cdots, \underline{f}^R, \overline{f}^{R+1}, \cdots, \overline{f}^M], \ a_k^i \equiv \overline{a}_k^i, \ b_k^i \equiv \overline{b}_k^i . \tag{4.11}$$

The values of the boundaries L and R are given by the Type-Reduction algorithm, previously presented in the Chap. 2.

4.2.2 Development of the n-step Ahead Predictor

When the developed model is integrated on a Model Predictive Control framework, the accuracy of the system's predicted behavior given by a locally linear model ultimately defines the success of the control law in accomplishing the desired closed loop performance metrics. However, the use of locally linear models to approximate a non-linear system's response several steps ahead of current sampling instant, particularly during transient conditions, is usually a sub-optimal approach. Nevertheless, in modeling and control applications, locally linear approximations usually yield computationally efficient methods without significant loss in accuracy [9].

Literature highlights two different ways of using locally linearized models to predict the behavior of a system during a future time window [9]:

- By considering a fixed linearized model constant over the prediction window (obtained at the extrapolation instant);
- By performing successive linearizations of the model at each new expected operating point over the prediction window.

While potentially more accurate, the latter approach has a larger computational burden directly dependent on the length of the prediction window. In addition, in mildly non-linear scenarios, when such approach is integrated in a closed loop control algorithm, it is known for not yielding significant improvements comparing to the former method [9]. For this reason, the present work will focus on the use of locally linear models which are assumed as constant over the whole prediction window.

Analyzing the simpler case based on a linearized Type-1 TS fuzzy model, one can obtain an approximation of the system's predicted response by evaluating its free response, assuming that future control actions will remain equal to the current control action $u(k)$ and disturbances ε are constant. Therefore, considering a one step-ahead predictor for a second order system given by Eq. (4.2) and assuming its dead-time equal to 0, two consecutive iterations of the predictor can be written as:

$$\hat{y}(k) = -a_1 y(k-1) - a_2 y(k-2) + b_1 u(k-d-1) + b_2 u(k-2) + \varepsilon(k)\ , \tag{4.12a}$$

$$\hat{y}(k+1) = -a_1 y(k) - a_2 y(k-1) + b_1 u(k) + b_2 u(k-1) + \varepsilon(k+1)\ . \tag{4.12b}$$

Thus, an incremental model can be obtained by subtracting two consecutive iterations, yields:

$$\begin{aligned} \hat{y}(k+1) = (1-a_1)y(k) - (a_2-a_1)y(k-1) + a_2 y(k-2) \\ + b_1 \Delta u(t) + b_2 \Delta u(t-1) + \varepsilon(k+1) - \varepsilon(k)\ , \end{aligned} \tag{4.13}$$

where $\Delta = 1 - z^{-1}$. Considering that a constant disturbance is present over the prediction window (N_p), the developed predictor will be offset-free (since $\varepsilon(k+1) - \varepsilon(k) = 0$). Thus, the free response (f(k+n)) over the future *n-steps* is recursively given by:

$$
\begin{aligned}
f(k+1) &= (1-a_1)y(k) - (a_2 - a_1)y(k-1) + a_2 y(k-2) \\
&\quad + b_1 \Delta u(k) + b_2 \Delta u(k-1) , \\
f(k+2) &= (1-a_1) f(k+1) - (a_2 - a_1)y(k) + a_2 y(k-1) + b_2 \Delta u(k-1) , \\
f(k+3) &= (1-a_1) f(k+2) - (a_2 - a_1) f(k+1) + a_2 y(k) , \\
f(k+4) &= (1-a_1) f(k+3) - (a_2 - a_1) f(k+2) + a_2 f(k+1) , \\
f(k+N_p) &= (1-a_1) f(k+N_p-1) - (a_2 - a_1) f(k+N_p-2) + a_2 f(k+N_p-3) .
\end{aligned} \tag{4.14}
$$

Regarding the application of the extrapolation hereby presented to Interval Type-2 TS Fuzzy Models, the obtained results remain valid by developing a *n-step* ahead predictor considering the parameters of the upper and lower bounds of the model separately and average the obtained estimations.

4.3 Application Scenarios

To assess the improvements attained with the use of IT2FLSs, two non-linear processes will be used as support for the development of the *n-step* ahead predictors. The Fermentation Reactor [9] and the Coupled Tanks systems [10] are two classical benchmark frameworks frequently used in the literature to evaluate the performance and robustness of non-linear modeling and control methodologies. To provide a comparative standpoint in the results discussion, two additional *n*-step ahead predictors will be implemented based on a linear ARX model and a Type-1 TS Fuzzy Model.

4.3.1 Fermentation Reactor Modeling

Yeast fermentation is a biochemical process that, having ethanol and carbon-dioxide as a sub-product, has significant value for several branches of food industry such as bakeries, breweries and distilleries as well as to other domains such as pharmaceutical and chemical plants. The yeast fermentation reaction is itself a composition of several interdependent physical/chemical processes and, for that reason, requires

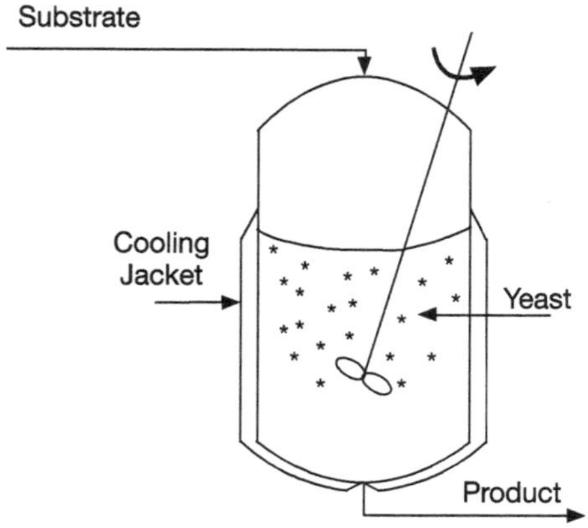

Fig. 4.1 Diagram of the continuous fermentation reactor

fairly complex models to be accurately simulated. In the work of [11], an extended model of this reaction is developed by complementing its kinetic properties (from which most part of the models existent in literature are based on [12]) with the heat transfer equations that directly influence the fermentation process. The biochemical reactions occurs in a reactor which is modeled as a stirred tank with constant substrate feed flow and a constant outlet flow containing the product (ethanol), substrate (glucose) and biomass (suspension of yeast), as generically depicted in Fig. 4.1.

The first-principle model of the yeast fermentation process is defined by the following set of non-linear differential equations:

$$
\begin{aligned}
\frac{dV(t)}{dt} &= F_i t - F_e(t) , \\
\frac{dc_X(t)}{dt} &= \mu_X(t) c_X(t) \frac{c_S(t)}{K_S + c_S(t)} \exp(-K_P c_P(t)) - \frac{F_e(t)}{V(t)} c_X(t) , \\
\frac{c_P(t)}{dt} &= \mu_P c_X(t) \frac{c_S(t)}{K_{S_1} + c_S(t)} \exp(-K_P c_P(t)) , \\
\frac{c_s(t)}{dt} &= -\frac{1}{R_{SX}} \mu_X(t) c_X(t) \frac{c_s(t)}{K_{S_1} + c_S(t)} \exp(-K_P c_P(t)) \\
&\quad - \frac{1}{R_{SP}} \mu_P c_X(t) \frac{c_S(t)}{K_{S_1} + c_S(t)} \exp(-K_{P_1} c_P(t)) \\
&\quad + \frac{F_i(t)}{V(t)} c_{S,in}(t) - \frac{F_e(t)}{V(t)} c_S(t) , \\
\frac{dc_{O_2}}{dt} &= k_{la}(t)(c^*_{O_2}(t) - c_{O_2}(t)) - \tau_{O_2}(t) - \frac{F_e(t)}{V(t)} c_{O_2}(t) ,
\end{aligned}
$$

$$\frac{dT_r(t)}{dt} = \frac{F_i(t)}{V(t)}(T_{in}(t)+273) - \frac{F_e(t)}{V(t)}(T_r(t)+273) + \frac{\tau_{O_2}(t)\Delta H_r}{32\rho_r C_{heat,r}}$$
$$- \frac{K_T A_T (T_r(t) - T_{ag}(t))}{V\rho_r C_{heat,r}} ,$$
$$\frac{dT_{ag}(t)}{dt} = \frac{F_{ag}(t)}{V_j}(T_{in,a} - T_{ag}(t)) + \frac{K_T A_T (T_r(t) - T_{ag}(t))}{V_j \rho_{ag} C_{heat,ag}} , \tag{4.15}$$

where

$$c^*_{O_2} = (14.6 - 0.3943T_r(t) + 0.007714T_r^2(t) - 0.0000646T_r^3(t)) \times 10^{-\sum(H_i I_i)(t)} ,$$
$$(H_i I_i)(t) = 0.5H_{Na}\frac{M_{Na}}{V(t)} + 2H_{Ca}\frac{M_{Ca}}{V(t)} + 2H_{Mg}\frac{m_{MgCl_2}}{M_{MgCl_2}}\frac{M_{Mg}}{V(t)}$$
$$+ 0.5H_{Cl}\left(\frac{m_{NaCl}}{M_{NaCl}} + 2\frac{m_{MgCl_2}}{M_{MgCl2}}\right)\frac{M_{Cl}}{V(t)} + 2H_{CO3}\frac{m_{CaCO3}}{M_{CaCO3}}\frac{M_{CO3}}{V(t)}$$
$$+ 0.5H_H 10^{-pH(t)} + 0.5H_{OH}10^{-(14-pH(t))} ,$$
$$k_{la}(t) = k_{la0}1.024^{Tr(t)-20} ,$$
$$\tau_{O_2}(t) = \mu_{O_2}\frac{1}{Y_{O_2}}c_X(t)\frac{c_{O_2}(t)}{K_{O_2} + c_{O_2}(t)} ,$$
$$\mu_X = A_1 \exp(-\frac{E_{a_1}}{R(T_r(t)+273)}) - A_2 \exp(-\frac{E_{a_2}}{R(T_r(t)+273)}) . \tag{4.16}$$

The model parameters and the nominal operation point of the system are presented in Tables 4.1 and 4.2, respectively.

Fermentation reactions are of exothermic nature and, since they are dependent on living organisms whose growth rate is highly sensitive to temperature variations, it is important to avoid temperature runaway of the reactor. Driven by this, temperature control is a key factor to ensure the reaction stability and can be efficiently used to indirectly obtain its sought products according to the demanded specifications. For this purpose and since sterility is often a crucial factor in such reactions, cooling jackets are usually employed as opposed to cooling coils into the fermenter itself [12]. Hence, from the perspective of a control algorithm, the reactor is a single-input single-output process: the coolant flow rate (F_{ag}) is the input (the manipulated variable) and the reactor's temperature (T_r) is the output (the controlled variable), as depicted in Fig. 4.2.

The dynamic behavior of the $T_r(F_{ag})$ dependency ultimately defines the complexity developed model. In the present scenario, both steady-state and dynamic properties of the process are of non-linear nature. As is shown by Figs. 4.3 and 4.4, the steady-state gain and the incremental gain are highly dependent on the current operating point.

Table 4.1 Parameters of the first-principle model of the yeast fermentation reactor

$A_1 = 9.5 \times 10^8$	$k_{la0} = 381\,h^{-1}$	$M_{Mg} = 24\,g\,mol^{-1}$
$A_2 = 2.55 \times 10^{33}$	$K_{la0} = 8.86\,mg\,l^{-1}$	$M_{MgCl_2} = 95\,g\,mol^{-1}$
$A_T = 1\,m^2$	$K_{la0} = 0.139\,g\,l^{-1}$	$M_{Na} = 23\,g\,mol^{-1}$
$C_{heat,ag} = 4.18\,J\,(g\,K)^{-1}$	$K_{la0} = 0.07\,g\,l^{-1}$	$M_{NaCl} = 58.5\,g\,mol^{-1}$
$C_{heat,r} = 4.18\,J\,(g\,K)^{-1}$	$K_{la0} = 1.03\,g\,l^{-1}$	$R = 8.31\,J\,(mol\,K)^{-1}$
$E_{a_1} = 55000\,J\,mol^{-1}$	$K_{la0} = 1.68\,g\,l^{-1}$	$R_{SP} = 0.435$
$E_{a_2} = 229999\,J\,mol^{-1}$	$K_{la0} = 3.6 \times 10^5\,J\,(h\,m^2\,K)^{-1}$	$R_{SX} = 0.607$
$H_{Ca} = -0.303$	$m_{CaCO_3} = 100\,g$	$V_j = 50\,l$
$H_{Cl} = 0.844$	$m_{MgCl_2} = 100\,g$	$Y_{O2} = 0.97\,mg^{-1}$
$H_{CO_3} = 0.485$	$m_{NaCl} = 500\,g$	$\Delta H_r = 518\,kJ\,(mol\,O_2)^{-1}$
$H_H = -0.774$	$M_{Ca} = 40\,g\,mol^{-1}$	$\mu_{O_2} = 241\,h^{-1}$
$H_{Mg} = -0.314$	$M_{CaCO_3} = 90\,g\,mol^{-1}$	$\mu_P = 241\,h^{-1}$
$H_{Na} = -0.550$	$M_{Cl} = 35.5\,g\,mol^{-1}$	$\rho_{ag} = 24\,g\,l^{-1}$
$H_{OH} = 0.941$	$M_{CO_3} = 60\,g\,mol^{-1}$	$\rho_r = 24\,g\,l^{-1}$

Table 4.2 Nominal operating point of the yeast fermentation reactor

$c_{O_2} = 3.106953\,mg\,l^{-1}$	$F_i = 51\,l\,h^{-1}$
$c_P = 12.515241\,g\,l^{-1}$	$pH = 6$
$c_S = 29.738924\,g\,l^{-1}$	$T_{ag} = 27.053939\,°C$
$c_{S,in} = 60\,g\,l^{-1}$	$T_{in} = 25\,°C$
$C_X = 0.904677\,g\,l^{-1}$	$T_{in,ag} = 20\,°C$
$F_{ag} = 18\,l\,h^{-1}$	$T_r = 29.4\,°C$
$E_{a_2} = 51\,l\,h^{-1}$	$V = 1000\,l$

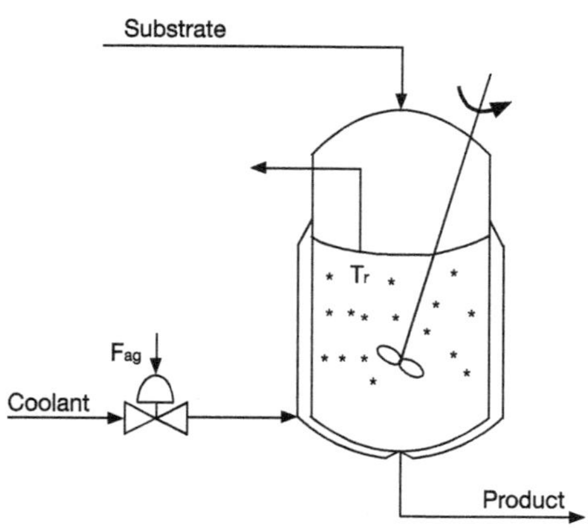

Fig. 4.2 Diagram of the continuous fermentation reactor considering the actuation and controller variables

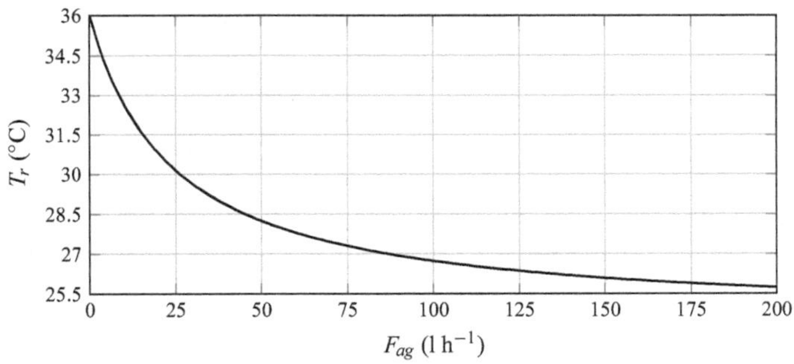

Fig. 4.3 Steady-state gain $T_r(F_{ag})$ of the yeast fermentation reactor

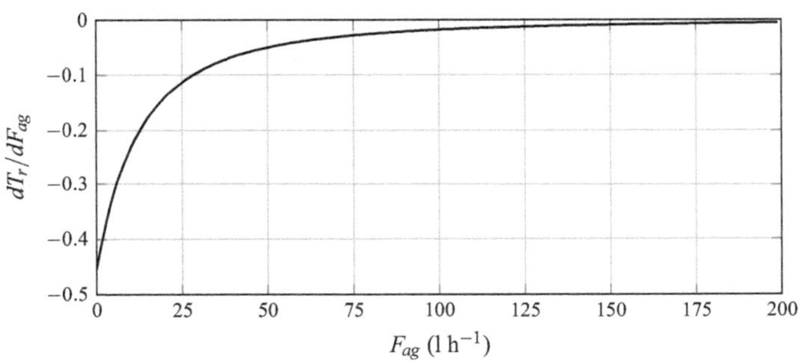

Fig. 4.4 Dependency of the incremental gain with the operating point

Yeast fermentation is a nonlinear process with relatively slow dynamic behavior which is mainly imposed by the glucose decomposition rate [11]. Consequently, when the reaction's operation point is changed, the attained settling time is in the scale of hours and, as so, one sample per hour is enough to capture the process's relevant dynamics.

In order to develop a model with "good" approximation capabilities over all the operational regimes of the process, a pseudo random sequence of control inputs was generated over the [0, 200] l h^{-1} control range, with steps changing every 200 h so the system reaches its steady-state. To better simulate the conditions of a real world scenario, the actuation variable was corrupted by gaussian noise with zero mean and variance of 0.1 l h^{-1} while the reactor's temperature measurements were corrupted by gaussian noise with zero mean and 0.05 °C variance. In Fig. 4.5, the training sequence and the system response are partially depicted (comprising a segment 2000 h of the complete training set), evincing the slow dynamics of the reactor's temperature variations.

On related literature [9], the same model is also used as a benchmark system and it was found that a second order regressive models with no dead-time are typically used to model this system. Accordingly, the regression variables considered during the identification of the model's consequent part parameters are following presented:

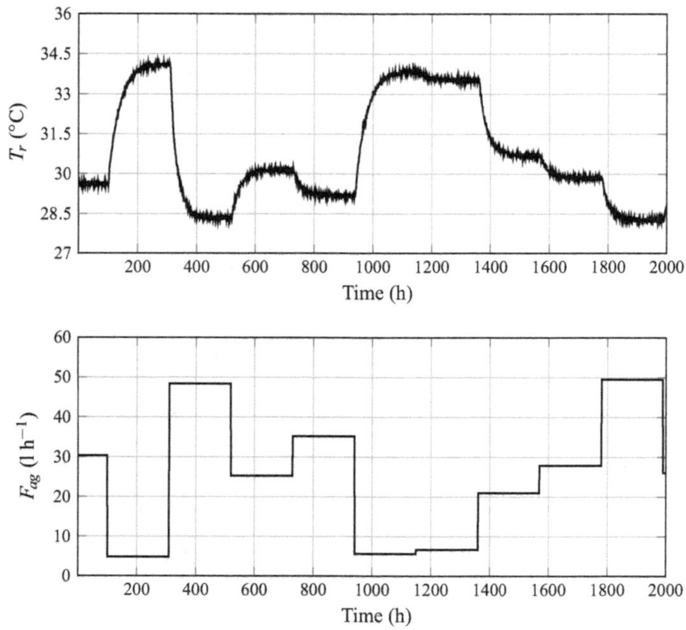

Fig. 4.5 Partial depiction of the response of the system to a pseudo-random control sequence used for model extraction

$$\hat{y}(k) = f(y(k-1), y(k-2), u(k-1), u(k-2)) . \tag{4.17}$$

The Subtractive Clustering algorithm was used to partition the model's inputs space thus obtaining the centers and variance of the membership functions that define a Type-1 TS Fuzzy Model. This process considers solely the $y(k-1)$ and $u(k-1)$ regressors as input variables, since small changes are observed over subsequent samples. The optimal input space partition was found considering yielding a total of 5 rules. The parameters of the membership functions that define the antecedent part of the model are presented in Table 4.3. Subsequently the consequent part parameters of the model were trained using the Recursive Least Squares procedure. The obtained Type-1 TS Fuzzy Model parameters will be then used as initialization of the Type-2 counterpart.

During the development of the Interval Type-2 TS Model its was found that its output is more sensitive to the width of the uncertainty interval of the consequent parameters than the antecedent ones. For this reason, and considering the conclusions obtained in Chap. 2 regarding the width of the antecedent Type-2 MF, a 5% uncertainty over the antecedent function's center was assumed, yielding a good coverage of the model's input space. Regarding the choice of the uncertainty interval of the consequent part parameters, several trials where performed to evaluate the influence of uncertainty ratio over the nominal value of the parameters. For each trial, it was considered that the system is at the steady-state operating point (where $F_{ag} = 32.5$

Table 4.3 Center and variance of each MF used in the Type-1 TS Fuzzy model input space partition

	$F_{ag}(\mathrm{l\,h^{-1}})$		$T_r(°\mathrm{C})$	
	$u(k-1)$		$y(k-1)$	
	Center	σ^2	Center	σ^2
Rule 1	5.6	15.36	33.28	3.78
Rule 2	26.1	15.36	30.15	3.78
Rule 3	58.3	15.36	29.92	3.78
Rule 4	88.3	15.36	27.01	3.78
Rule 5	119.6	15.36	26.43	3.78

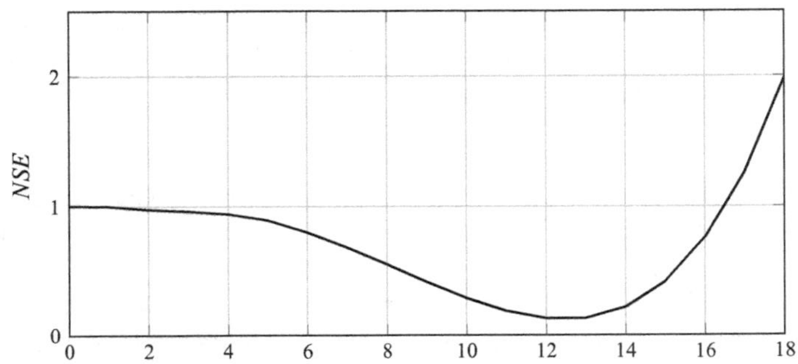

Fig. 4.6 Normalized Squared Error of 10 step-ahead estimations using a Type-2 TS Fuzzy model for different uncertainty ratios over the consequent part parameters, normalized to the Type-1 TS Fuzzy model estimation error

$\mathrm{l\,h^{-1}}$ and $T_r = 29.4$ °C), performed then a step in the control signal (F_{ag}) in the range $\Delta u(k) \in [-32.5, 37.5]$. For each step, the change in the process output (T_r) is then obtained 10 sampling instants after. Due to the change of the process gain over different operation points, it is expected the model approximation capabilities to deteriorate when large changes in the operation point are performed. Therefore, long prediction horizons may reveal inadequate since the local linearization used in the extrapolation may no longer be valid. Figure 4.6 illustrates this procedure, revealing the influence of the uncertainty bounds in the prediction error of the Type-2 TS model.

As is clear from the obtained results, which are normalized to the error obtained with a Type-1 TS model, increasing the uncertainty width of the Type-2 TS model's parameters is beneficial for the model prediction capabilities. This behavior is observed approximately up to 12% uncertainty ratio, from which its prediction capabilities start to deteriorate. Once again, the results obtained for the 0% value (NSE = 1) supports the equivalence between a Type-1 and a Type-2 TS model with 0% uncertainty ratio.

In Figs. 4.7, 4.8 and 4.9, are presented the Type-2 TS model 10 step-ahead predictions for several changes in the control signal in the conditions of the previous

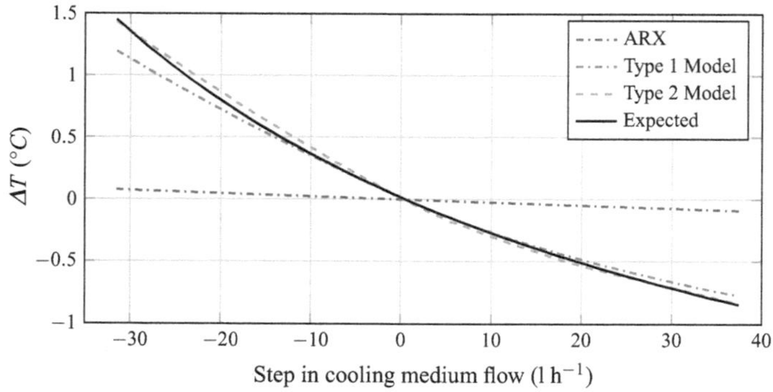

Fig. 4.7 Reactor's temperature variation for different actuation steps over the nominal point (29.4 °C). The 10 step-ahead estimations given by predictors based on a linear ARX model, Type-1 TS Fuzzy model and a Type-2 TS Fuzzy model with 12% uncertainty ratio over the consequent part parameters

evaluation considering the 12, 5 and 15% uncertainty ratios. These results are then compared with extrapolation obtained by two additional predictors (ARX and Type-1 TS FLS).

As is observed in Fig. 4.7, the use of a single linear ARX predictor clearly reveals an approach incapable of providing long term predictions due to the significant change of the process gain. On the contrary, for the given scenario, the TS Fuzzy Models reveal their superior interpolation capabilities by providing better predictions of the local behavior of the system over the considered horizon. The results obtained also show that the Type-1 TS model predicts the system behavior with a small error relatively to the real system behavior. This occurs when the step in the cooling medium is limited to the interval $\Delta u(k) \in [-10, 10]$. As the step amplitude increases, the model starts to diverge. When a Type-2 TS model with an uncertainty factor of 12% over the consequent parameters is used, the model remains close to the real behavior of the system over the whole considered interval. As will be shown in the Chap. 5 this improvement is important for the performance enhancements of the predictive controller developed based the Type-2 TS Fuzzy Model framework. In Figs. 4.8 and 4.9, the influence of the uncertainty interval width at the consequent part parameters is evaluated for 5 and 15% scenarios.

In the former figure the differences between the Type-1 and Type-2 TS Fuzzy Models are little (due to the small width of the considered intervals and consequent equivalence of the models). In contrast, when the uncertainty width is increased significantly, the Type-2 Fuzzy Model is no longer advantageous, as is evinced in the latter case where it presents a larger error over the prediction horizon than its Type-1 counterpart. While in this scenario the magnitude of the error is relatively small, the increase of the model's parameters uncertainty results in an increase of the predictor's error in the vicinity of the current operating point (for small variations of the cooling medium flow rate). When such model is integrated in a closed loop model-based controller and operating at steady-state, such predictor inaccuracies will ultimately produce a more active control signal—which is not necessarily desirable.

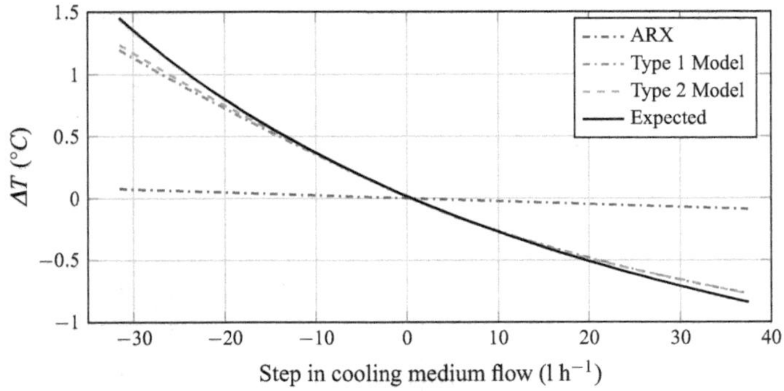

Fig. 4.8 Reactor's temperature variation for different actuation steps over the nominal point (29.4 °C). The 10 step-ahead estimations given by predictors based on a linear ARX Model, Type-1 TS Fuzzy model and a Type-2 TS Fuzzy Model with 5% uncertainty ratio over the consequent part parameters

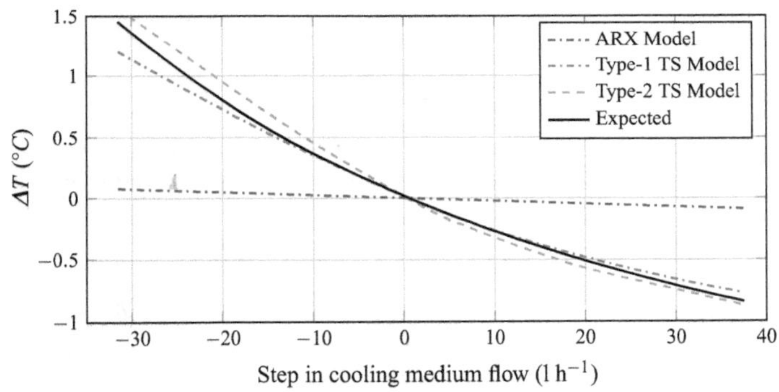

Fig. 4.9 Reactor's temperature variation for different actuation steps over the nominal point (29.4 °C). The 10 step-ahead estimations given by predictors based on a linear ARX Model, Type-1 TS Fuzzy model and a Type-2 TS Fuzzy Model with 15% uncertainty ratio over the consequent part parameters

4.3.2 *Coupled Tanks Modeling*

The control of the liquid level in tanks is a common problem in industries that require fluids to be pumped and stored between several deposits. Often, such tanks are used to perform mixing reactions that, depending on their nature, can result in sudden volumetric changes of the liquids, thus requiring a tight control of their level and flow rates to comply with the tank's storage capacities. The Coupled Tanks System (CTS) [10] used in this evaluation is based on a small scale system consists of two tanks, having each one an independent pump to control the inflow of liquid and an outlet at the bottom responsible for the liquid leakage. Additionally, the tanks are interconnected by a channel which allows the liquid to flow between them. A diagram of this setup is presented in Fig. 4.10.

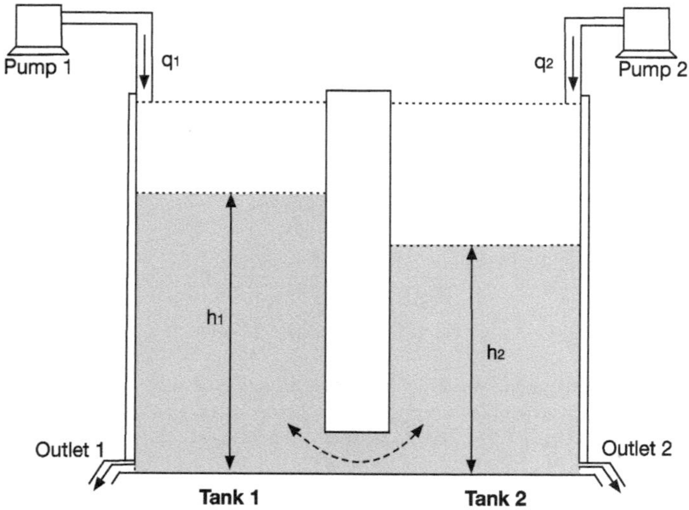

Fig. 4.10 Diagram of the coupled tanks system

Based on the Bernoulli's equations for a non-viscous, incompressible fluid in steady-state flow, the dynamics of the CTS can be modeled by the resulting set of non-linear differential equations:

$$a_1 \frac{dh_1}{dt} = q_1 - \alpha_1 \sqrt{h_1} - \text{sgn}(h_1 - h_2)\alpha_3 \sqrt{|h_1 - h_2|}\,, \tag{4.18a}$$

$$a_2 \frac{dh_2}{dt} = q_2 - \alpha_2 \sqrt{h_2} + \text{sgn}(h_1 - h_2)\alpha_3 \sqrt{|h_1 - h_2|}\,. \tag{4.18b}$$

where a_1 and a_2 denote the cross-sectional area of the tank 1 and 2, h_1 and h_2 are the liquid level in tank 1 and 2, q_1 and q_2 are the volumetric flow rate (cm^3 s^{-1}) of Pump 1 and 2, α_1, α_2 and α_3 are proportionality coefficient corresponding to the $\sqrt{h_1}$, $\sqrt{h_2}$ and $\sqrt{h_1 - h_2}$ terms which depend on the discharge coefficients of each outlet and the gravitational constant. The reservoir model parameters were obtained from the setup described in [13], and are presented in Table 4.4.

By choosing which pumps and rotary valves are directly manipulated, one can develop either a Single-Input Single-Output (SISO) or a Multiple-Input Multiple-Output (MIMO) system model to evaluate the variations of the liquid level of the

Table 4.4 Parameters of the simulated coupled tanks system

a_1	a_2	α_1	α_2	α_3
36.52 cm^2	36.52 cm^2	5.6186	5.6182	10

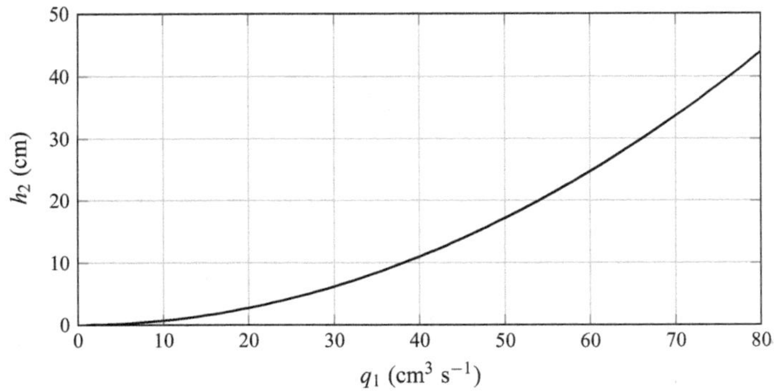

Fig. 4.11 Steady-state gain $h_2(q_1)$ of the coupled tanks system

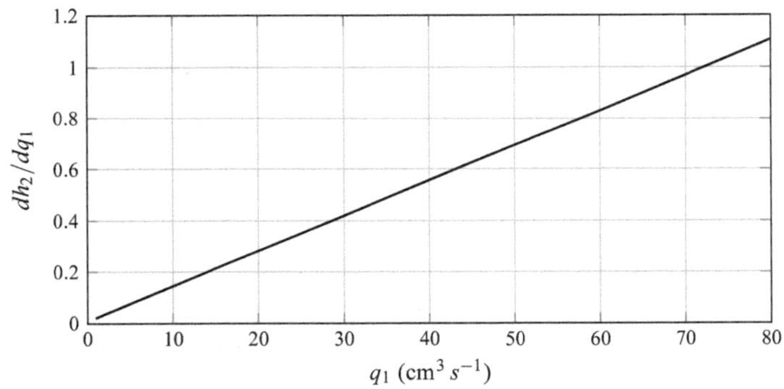

Fig. 4.12 Dependence of the incremental gain with the operating point

tanks. In the present evaluation scenario, the Pump 1 will be the actuation variable and the liquid level of the Tank 2 the controlled one and is considered that the valves at the tank interconnection and respective outlets maintain a constant aperture. Using this configuration, the presented setup can be considered as a non-linear second order SISO system.

On related literature [10], this system is modeled using second order regressive models with no dead-time. Accordingly, the regression variables considered during the identification of the model's consequent part parameters are:

$$\hat{y}(k) = f(y(k-1), y(k-2), u(k-1), u(k-2)) \,. \tag{4.19}$$

As is evinced by Figs. 4.11 and 4.12 the system's steady-state gain is of non-linear nature and its incremental gain highly dependent on the current operation point.

In order to extract a model capable of approximate the system's operation region, a pseudo random sequence of control inputs was generated (considering the maximum flow rate for the Pump 1 to be 80 cm^{3}s^{-1}), with steps changing every 200 s and the Pump 2 turned off. By evaluating several step responses, a sampling interval of 2 s was chosen as appropriate to capture the plant's behavior. An unmeasured

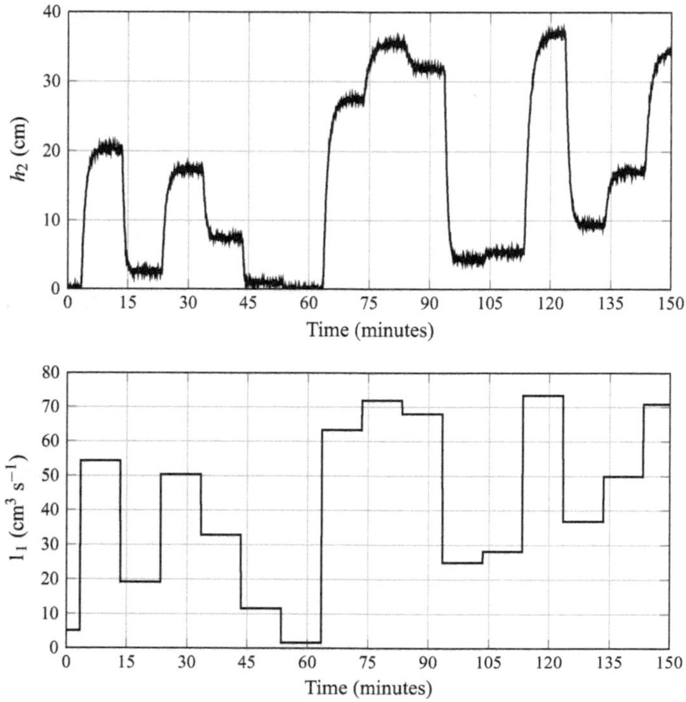

Fig. 4.13 Response of the system to a pseudo-random control sequence used for model extraction

gaussian disturbance is introduced in the control signal of Pump 1 with zero mean and variance of $0.5\,\text{cm}^3\,\text{s}^{-1}$ while the Tank 2 liquid level measurements are corrupted by a gaussian noise with zero mean and variance of 0.05 cm. Figure 4.13 partially details the changes in the Tank 2 level to a sequence of different liquid flow rates at the Tank 1 inlet.

Once again, for comparative purposes, three models were developed based on the linear ARX, Type-1 and Type-2 TS Fuzzy Models. The Type-1 TS Fuzzy model was obtained using the Subtractive Clustering algorithm considering solely the $y(k-1)$ and $u(k-1)$, followed by a training of the consequent part parameters based on the Recursive Least Squares Algorithm. The optimal input space partition was found yielding a total of 4 rules, which come out to the antecedent part parameters presented in Table 4.5.

Similarly to the previous scenario, the influence of the uncertainty ratio ration over the Type-2 TS model parameters was evaluated by measuring the model's 10 step-ahead prediction error considering the system at the steady-state, with $q_1 = 60\,\text{cm}^3\text{s}^{-1}$ and $h_2 = 24.7\,\text{cm}$ and control step variations in the interval $\Delta u \in [-50, 20]$. The Type-2 MF at the antecedent part of the rule base were obtained considering a 5% uncertainty ratio over the Type-1 ones. The obtained results are represented in Fig. 4.14.

Table 4.5 Center and variance of each MF used in the Type-1 TS Fuzzy model input space partition

	$q_1(cm^3\ s^{-1}$		h_2(cm)	
	$u(k-1)$		$y(k-1)$	
	Center	σ^2	Center	σ^2
Rule 1	8.6	11.30	0.79	6.20
Rule 2	32.8	11.30	7.51	6.20
Rule 3	54.1	11.30	19.92	6.20
Rule 4	71.9	11.30	35.06	6.20

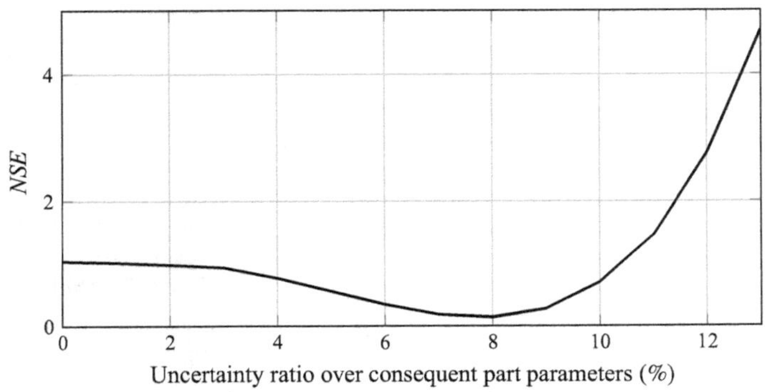

Fig. 4.14 Normalized squared error of 10 step-ahead estimations using a Type-2 TS Fuzzy model for different uncertainty ratio over the consequent part parameters, normalized to the Type-1 TS Fuzzy model estimation error

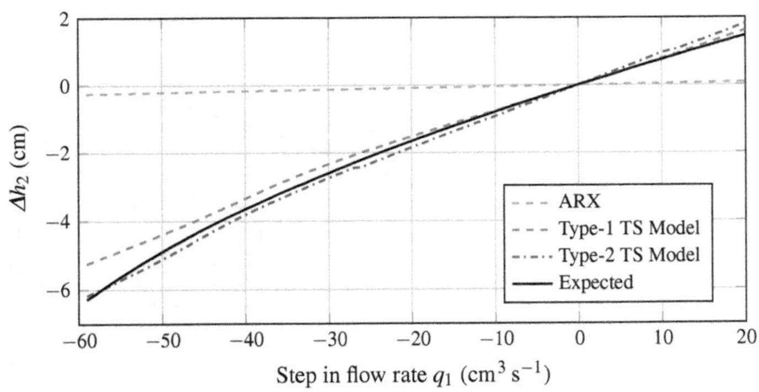

Fig. 4.15 Tank 2 liquid level variation (nominal value of 24.7 cm) due to a actuation steps in the Pump 1. Comparison between the estimations given by a predictor based on a linear ARX Model, Type-1 TS Fuzzy model and a Type-2 TS Fuzzy Model with 8% uncertainty width over the nominal consequent part parameters

In comparison to the previous evaluation scenario, the uncertainty ratio introduced in the consequent part parameters has a similar influence over the prediction model's accuracy, showing that increasing its value is beneficial up to a certain limit. In the Coupled Tanks model case, the optimal Type-2 TS model is obtained when a uncertainty ratio of 8% is used over the nominal parameters' values.

Figures 4.15, 4.16 and 4.17, for Type-2 TS models results are obtained considering uncertainty widths of 8%, 4%, and 12% respectively, depict the prediction accuracy of the three evaluated models for different step changes over the nominal operation point.

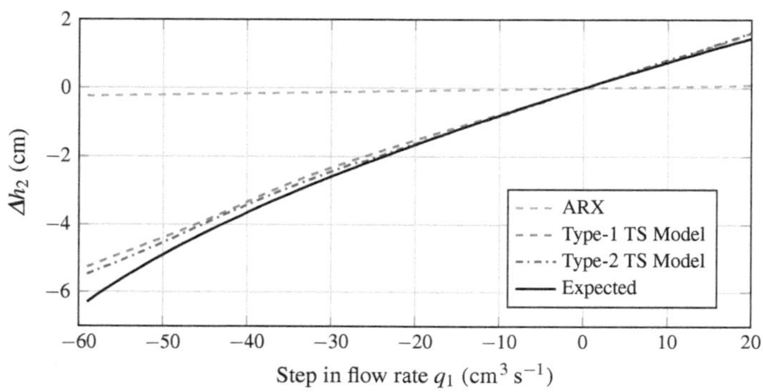

Fig. 4.16 Tank 2 liquid level variation (nominal value of 4.7 cm) due to a actuation steps on the Pump 1. Comparison between the estimations given by a predictor based on a linear ARX model, Type-1 TS Fuzzy Model and a Type-2 TS Fuzzy Model with 4% uncertainty width over the nominal consequent part parameters

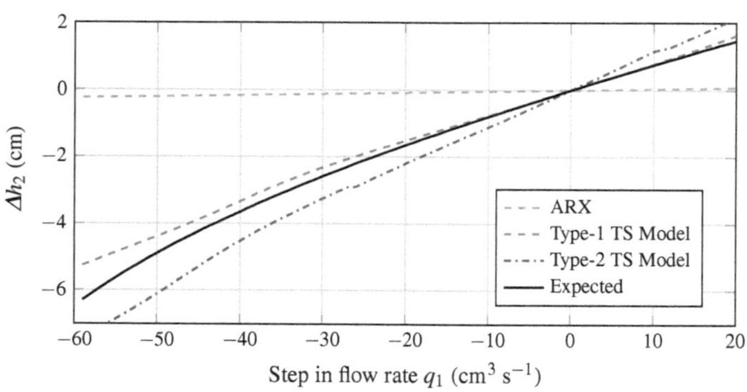

Fig. 4.17 Tank 2 liquid level variation (nominal value of 24.7 cm) due to a actuation steps in the Pump 1. Comparison between the estimations given by a predictor based on a linear ARX model, Type-1 TS Fuzzy model and a Type-2 TS Fuzzy model with 12% uncertainty width over the nominal consequent part parameters

Similarly to the previously evaluated scenario, it is clear that the considered uncertainty widths in the Type-2 TS Fuzzy Model significantly influence the performance of the developed predictor. The scenario with 8% uncertainty width presented the best results and for that reason, will be used in the controller synthesis further presented.

4.4 Conclusions

In the present chapter an approach for the development of a predictive Type-2 TS Fuzzy Model based on the weighted contribution of locally linear models was presented. The evaluated scenarios show that the uncertainty width considered in the Type-2 TS Fuzzy Model parameters significantly influence the developed predictor's performance and, for this reason, the choice of this parameter may be subject to an optimization procedure. One relevant conclusion is that the prediction model's performance can be slightly improved based on the Type-2 FL, without increasing the model complexity with new rules and regression parameters thus providing a better support for model-based linear controllers. The succeeding chapter will develop such application, providing comparative results of a closed loop control system based on the described non-linear processes.

References

1. Iserman, M., Münchhof, M.: Identification of Dynamic Systems. Springer, Heidelberg (2011)
2. Nørgaard, M., Ravn, O., Poulsen, K., Hansen, L.: Neural Networks for Modeling and Control of Dynamic Systems. Springer, London (2000)
3. Ljung, L.: System Identification: Theory for the User. Prentice Hall (1987)
4. Åström, K.; Wittenmark, B.: Adaptive Control. Dover, London (2008)
5. Clarke, D., Mohtadi, C., Tuffs, P.: Generalized predictive control, parts 1 and 2. Autom. **23**, 137–160 (1987)
6. Herceg, M., Kvasnica, M., Fikar, M.: Transformation of Fuzzy Takagi-Sugeno Models into Piecewise Affine Models. Rough Sets and Intelligent Systems Paradigms, Lecture Notes in Computer Science **4585**, 211–220 (2007)
7. Mendes, J.: Computational Intelligence Methodologies for Control of Industrial Processes, Ph.D. Thesis, Department of Electrical and Computer Engineering, University of Coimbra (2014)
8. Bishop, C.: Pattern recognition and Machine Learning, Information Sciences and Statistics, 1st edn. Springer, New York (2006)
9. Ławryńczuk, M.: Computationally Efficient Model Predictive Control Algorithms – A Neural Network Approach, Studies in Systems, Decision and Control, vol. 3. Springer (2014)
10. Ogata, K.: Modern Control Engineering, 5th edn. Prentice Hall (2009)
11. Nagy, Z.: Model based control of a yeast fermentation bioreactor using optimally designed artificial neural networks. Chemical Engineering Journal, **127**: 95–109. Elsevier(2007)
12. Luyben, W.: Chemical Reactor Design and Control. Wiley, New York (2007)
13. Wu, D., Tan, W.: A simplified Type-2 Fuzzy logic controller for real-time control. ISA Trans. **45**(4), 503–516 (2006)

Chapter 5
Model Predictive Control Using Type-2 Takagi-Sugeno Fuzzy Systems

5.1 Introduction

Model Predictive Control (MPC) is one of the most researched synthesis approaches which, based on a model of a process, computes the best control strategy according to a set of predefined goals over a future time horizon. In fact, this is one of the most distinguishable features of MPC. While, traditionally, control systems determine the course of actions based on the evolution of the error of previous iterations, MPC is driven by evaluating the expected future error due to a chosen control trajectory in a receding horizon fashion. The success of MPC is in part due to the following factors [1]:

- Applicability to a broad class of systems which are difficult to control (i.e. processes with significant time-delays or non-minimum phase behavior)
- Ability to handle constraints imposed in the control as well in the system states
- Algebraic approach to obtain a closed-loop controller
- Easily extended to MIMO processes
- Good tracking performance
- Computational feasibility

The genesis of MPC goes back to early 1980s [2] as a result from the commitment of practitioners to solve specialized control needs of power plants and petroleum refineries. Due to its origins, many of the MPC practical implementations are currently found in industry, mainly in scenarios where the time constants of the controlled processes are measured in the scale above seconds. Examples of its practical feasibility are found in chemical, petrochemical, paper or in food processing industries [2]. They represent applications where the control goals usually need to be stated not solely based on a reference signal (as in classical control systems such as PID or Pole-Placement), but also considering several natural evaluative metrics related with the quality of the final products and their cost.

Motivated by its high level of "simplicity" and flexibility in handling complex control scenarios, academia continued to put efforts in developing MPC beyond just

R. Antão, *Type-2 Fuzzy Logic*, Nonlinear Physical Science,
DOI 10.1007/978-981-10-4633-9_5

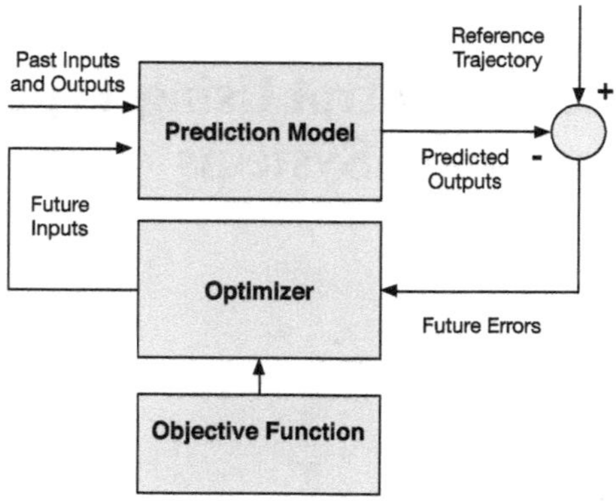

Fig. 5.1 Basic structure of a MPC algorithm

industrial practices, proposing several extensions based on the most recent developments in system's modeling and optimization domains to address complex non-linear problems in any potential controllable scenario [3]. While in the past the available computational resources restrained their use in real-time applications, the theoretical results achieved were not unfruitful. Nowadays, the available computational power even in small embedded systems is turning some of the proposed non-linear MPC approaches into successful practical applications. Dynamic systems with *fast* time constants in automotive industry [4], such as in engine control, or in aerospace applications, as in Unmanned Aerial Vehicle flight control system, [5] are some examples of its recent applications.

MPC is not an unique technique but rather a set of different methodologies rooted on three common functional blocks, interconnected as presented in Fig. 5.1.

Due to the predictive nature of this control approach, the model's process plays a central role in a MPC algorithm as it provides the mean to extrapolate the expected future behavior of the system. The quality of its predictions ultimately not only determine the capability of achieve the control goals, but also drive the level of computational complexity required to obtain the optimal control sequence. As was previously pointed, MPC algorithms were developed from practitioners and, for that reason, traditionally consider the simplest model capable of give accurate enough predictions. Linear models often satisfy this paradigm, assuming that their uncertainty and some gain scheduling in the control law suffices to overcome mildly non-linearities that industrial processes present [6]. Among the linear modeling approaches found in the related literature, one can distinguish three main groups:

- Truncated Impulse Response Models—traditionally favored by industry applications. These models are obtained by performing simple tests such as the step response and easily interpretable, but require far more data to tune their large number of parameters than the following two approaches

- State Space Models—most widespread in the academic research, particularly in the USA, allowing simple derivations for the controller and easily extended to the multi-variable case
- Transfer Function Models and polynomial methods—curiously these methods are preferred by academics in Europe, and clearly evince the influence of concepts such as dead-time and time constants in their representations. Furthermore, their structure can be easily constructed using both technological knowledge about the process and information provided by data-sets [3]. Due to the easiness in obtaining a regression model, such approach has a closer connection to popular black-box identification techniques what grants it an additional advantage in discrete-time process modeling. However, in multidimensional scenarios, transfer function models may lead to cumbersome and non minimal representations which are harder to interpret and manipulate

Nevertheless, to address physical processes with strong non linearities, literature proposes several models based on non-linear structures such as Cascade (serial) models, Volterra series, Wavelets, Neural Networks and ultimately Fuzzy Systems [3]. Among the referred approaches, non-linear MPC implementations tend to give preference to neural models, existing a wider consensus in the optimal approaches to use such structures in the predictive control domain [3]. This option is mainly due to their advantages in terms of approximation accuracy (they are well known universal approximators), reasonably low number of parameters and a simple structure. Moreover, a considerable number of training and structure optimization algorithms are available for neural models, which make the modeling task a seamless process for practitioners less experienced with the physical details and the relevant variables of the controlled process.

On the other hand, the concept of incorporating FL into neural models has become increasingly important in recent years and has been object of important studies regarding their stability and applicability in MPC [7]. In contrast to the pure neural networks and fuzzy systems, neuro-fuzzy models based on the Takagi–Sugeno structure have been proven suitable for the use in non-linear MPC. In [8–10] are presented some examples of the successful applications of MPC using fuzzy models.

Regarding the objective function, the first algorithm ever proposed according to the MPC principles, known as Model Algorithmic Control (MAC) [11], was designed to solely minimize the predicted deviations of the process from the reference trajectory in a least-squares sense. The Dynamic Matrix Control (DMC) [12] algorithm was in fact the first to use the two evaluative metrics that are nowadays considered as standard elements of a MPC objective function. The first one takes into account the differences between the predicted trajectory of the output variable and the set-point trajectory (i.e. the predicted control errors) over the prediction horizon N_p. The role of the second part of the objective function is to introduce a penalty term to reduce excessive (and hence disadvantageous) changes of the manipulated variable.

In addition, it also improves the numerical properties of the optimization process [6]. Therefore, the generic quadratic cost function is typically used:

$$J = \sum_{p=1}^{N}[\hat{y}(k+p|k) - r(k+p)]^2 + \sum_{p=1}^{N}[\Delta u(k+p-1|k)]^2 \,, \tag{5.1}$$

where $\hat{y}(k+p|k)$ is a p-step ahead predictor of the system on instant k, $r(k+p)$ is the future reference trajectory and $\Delta u(k+p-1|k)$ is the control signal increment and N is the optimization horizon (for the predictor and control signal in this case).

It is important to note that the relevance of each performance index to the final cost function can be adjusted according to several weighting factors. They are usually tuned to achieve a desired closed loop dynamic performance, but they can also be dictated by economic objectives of the control system. The choice of a well posed performance index is also a fundamental condition when developing a MPC strategy capable of achieve offset free reference tracking [6]. By other words, in steady state, the minimum of J must be consistent with zero tracking errors. Equation (5.1) satisfies this condition by considering Δu rather u in the cost function penalty factor. If the absolute value of u was used instead, the performance index would be biased, favoring operation points which require control signals with smaller absolute magnitude. For obvious reasons such constraint does not comply with the goals of a closed loop control system with a wide operation region.

Ultimately, the choices taken in the previous blocks sum up in an optimization problem that has to be solved online in between each sampling instant according to the new information inferred by the current state of the process. The reliance of MPC in linear models is motivated by the reduced computational complexity necessary to find the solution for the optimization problem and easiness in developing adaptive strategies to cope with time-varying conditions in specific application scenarios [13]. Although providing a sub-optimal representation of non-linear processes, linear models allow one to synthesize a simple linear algebra problem yielding an explicit solution of the problem. On the other hand, whereas non-linear models invariably escalate the complexity of synthesizing the control law to a non-linear optimization problem, which must be solved using heavy iterative methods such as the Levenberg-Marquardt [14, 15] or the Broyden–Fletcher–Goldfarb–Shanno [16] algorithms. Furthermore, such methods provide no guarantees to find an optimal solution in useful time due to the non-convex nature of the problem. In this perspective, under certain circumstances it is an acceptable compromise to work with a sub-optimal model. As will be further presented in this chapter, by using linear approximations of non-linear models in the vicinity of the current process operation point, one is capable to retain a significant modeling accuracy (ultimately defining the success in closed loop reference tracking) while simultaneously reducing the optimization problem complexity.

5.2 Generalized Predictive Control

One of the most used implementations of MPC in industry is the Generalized Predictive Control (GPC) [17] due to its simplicity, flexibility and robustness in controlling complex systems, where self-tuning controller methods such as pole-placement and other minimum-variance methods [13] are less efficient due to their sensitivity to initial design assumptions. Similarly to many other MPCMPC implementations, in its unconstrained form the GPC synthesizes the control law based on the minimization of a multi-step cost function that weights the quadratic terms of the control error and the control increments on a finite time horizon into the future, as represented:

$$J = \sum_{p=d+1}^{N_p} [\hat{y}(k+p|k) - r(k+p)]^2 + \sum_{p=d+1}^{Nu} [\lambda(z^{-1})\Delta u(k+p-1|k)]^2 , \quad (5.2)$$

where $\hat{y}(k+p|k)$ is a p-step ahead predictor of the system on instant k, $r(k+p)$ represents the future reference trajectory, $\Delta = 1 - z^{-1}$, $\lambda(z^{-1})$ stands for a weight polynomial introducing a penalty factor over the control signal activity and $d+1$ is the first output value that can be controlled from the present iteration. The variables N_p and N_u are the prediction and control horizons, respectively. Figure 5.2 illustrates the main GPC concept, considering a dead-time $d = 1$.

The choice of the parameter d is deeply related with the system dead-time. If a system has a dead-time of 2 samples, for example, it would be superfluous to try to minimize the difference $y(k+1) - r(k+1)$ since this quantity although in the

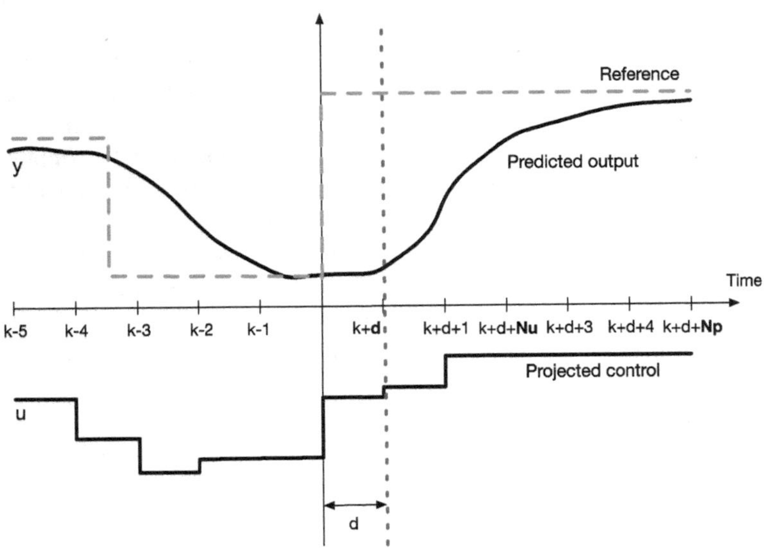

Fig. 5.2 Reference, control and predicted output in a GPC algorithm

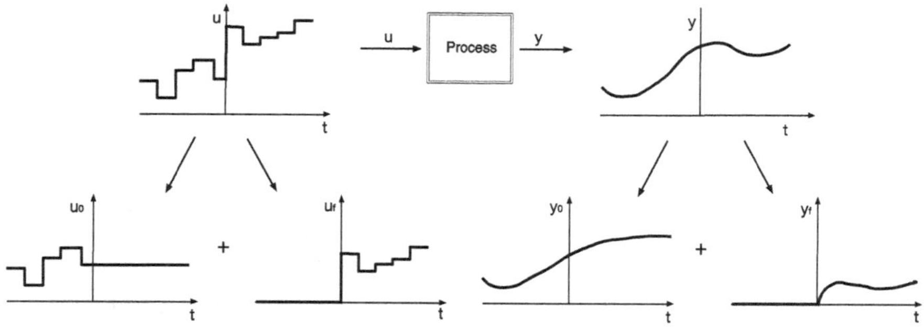

Fig. 5.3 Free and forced response

future, cannot be influenced by present or future control actions. When no *a priori* knowledge of the delay is present, it is usual to set this value to 0. Concerning the length of the prediction window (N_p), this value should be selected in order to ensure the predictor's validity over the chosen interval - if the process is of non-linear nature, its dynamics may change significantly during the prediction window. Regarding the parameter N_u, it is pointless to attribute a value higher than $(N_p - d - 1)$, since only control actions up to this point would have effect in the output within the horizon N_p. Nonetheless, as is suggested by [17] the value of N_u is usually substantially smaller than the prediction horizon. The concept of control horizon is in fact one of the novelty factors introduced by the GPC strategy, which considers the control signal increments null after the N_u window. This simplification significantly reduces the dimensionality of the matrix operations required to compute the control signal and improves the robustness of the control law when unknown future weighting factors, (λ), are required for the controller to be realizable.

When no other constraints are imposed to the process' variables, the attainment of the optimal control signal is a simple algebraic cost function minimization problem. Without entering at this stage in significant calculations, the GPC algorithm synthesis problem based on linear models starts by assuming that the predicted output of the model, $\hat{y}$ results form the contribution of the forced response of the system y_f and its free response y_0, as presented in Fig. 5.3. The forced response is determined by future increments of the manipulated variable while the free response depends only on the past values of the manipulated variable and the previous values of the system output.

Based on this property, one can rewrite the model transfer function as presented:

$$\hat{\mathbf{y}} = \mathbf{y}_f + \mathbf{y}_0 \,, \tag{5.3}$$

which can be further rewritten as

$$\hat{\mathbf{y}} = \mathbf{G}\Delta\mathbf{u} + \mathbf{y}_0 \,, \tag{5.4}$$

where G is a N_p by N_u lower triangular matrix representing the impulse response of the transfer function from u to y

$$\begin{bmatrix} g_{1,0} & 0 & 0 & \cdots & 0 \\ g_{2,1} & g_{2,0} & \cdots & \cdots & 0 \\ g_{3,2} & g_{3,1} & g_{3,0} & \cdots & 0 \\ \vdots & \vdots & \vdots & \ddots & \vdots \\ g_{N_p,N_p-1} & g_{N_p,N_p-2} & g_{N_p,N_p-3} & \cdots & g_{N_p,N_p-N_u} \end{bmatrix} . \tag{5.5}$$

It worths noting that if the plant dead-time is $d > 0$ the first p rows of G will be null. Generally in self-tunned modeling scenarios d will not be known *a priory* but, despite that, GPC theory provides a stable solution for the minimization problem even if the leading rows of G are zero, as long as a reasonable estimate of the model order is used [17].

Assuming that the control activity weight factor λ is constant over the time, the cost function (5.2) can be represented using a vector/matrix form, as follows:

$$\begin{aligned} J &= (\hat{\mathbf{y}} - \mathbf{r})^T(\hat{\mathbf{y}} - \mathbf{r}) + \lambda \Delta\mathbf{u}^T \Delta\mathbf{u} \\ &= (\mathbf{G}\Delta\mathbf{u} + \mathbf{y}_0 - \mathbf{r})^T(\mathbf{G}\Delta\mathbf{u} + \mathbf{y}_0 - \mathbf{r}) + \lambda\Delta\mathbf{u}^T\Delta\mathbf{u} . \end{aligned} \tag{5.6}$$

By setting the partial derivatives of J with respect to Δu as zero, one obtains the control law for the unconstrained scenario as presented in the equation:

$$\Delta\mathbf{u} = (\mathbf{G}^T\mathbf{G} + \lambda\mathbf{I})^{-1}\mathbf{G}^T(\mathbf{r} - \mathbf{y}_0) . \tag{5.7}$$

As GPC is used in a receding horizon fashion, only the first optimal control increment of $\Delta\,\mathbf{u}$ is applied and so, the effective control signal is obtained as $u(k) = u(k-1) + \Delta\mathbf{u}(1)$. At the next sampling event, the "optimal" control increment is calculated once again considering new N_p step-ahead predictions, effectively closing the control loop.

From Eq. (5.7), a very simple interpretation regarding the operation of the GPC can be attained. If the expected future reference tracking error is zero, i.e. the free response of the system y_0 is sufficient to achieve the control set-point r, then no further increments in the control signal are necessary. Otherwise, there will be an increment in the control action proportional to the future error independently from the current absolute value of the control variable. Moreover, it is evinced the importance of the prediction model accuracy: since the controller gain is solely proportional to the future error, if the model predictions become biased then the controller will not achieve the sought zero error reference tracking.

Figure 5.4 summarizes the procedure of synthesizing a control law based on the GPC theory.

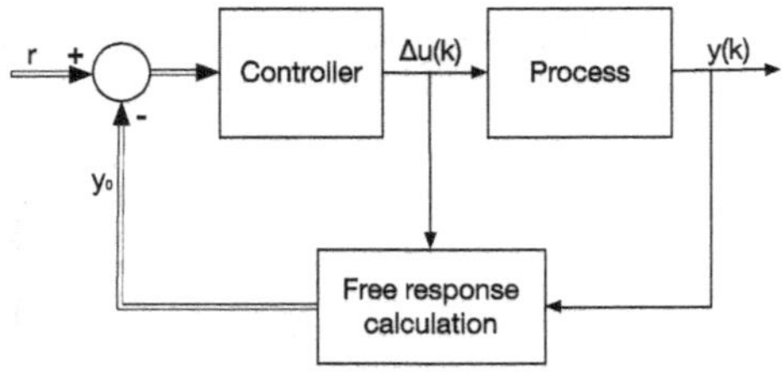

Fig. 5.4 GPC algorithm control

5.3 Derivation of a *n*-step Ahead Predictor

To optimize the cost function defined in Eq. (5.2) one has to firstly obtain an optimal prediction $\hat{y}$ for every sample within the prediction horizon. Most SISO plants can be described around a particular set-point and after linearization by a Controlled Auto-Regressive Moving Average (CARMA) model:

$$A(z^{-1})y(k) = B(z^{-1})z^{-d}u(k-1) + \frac{C(z^{-1})}{\Delta}v(k)\ , \tag{5.8}$$

where $u(k)$ and $y(k)$ are the input and output sequences of the plant, d is the dead time of the system, $v(k)$ is an unknown disturbance and $\Delta = 1 - z^{-1}$. The symbols A, B, and C denote the following polynomials in the backward shift operator z^{-1}:

$$\begin{aligned} A(z^{-1}) &= 1 + a_1 z^{-1} + a_2 z^{-2} + \cdots + a_{na} z^{-na}\ , \\ B(z^{-1}) &= b_1 z^{-1} + b_2 z^{-2} + \cdots + b_{nb} z^{-nb}\ , \\ C(z^{-1}) &= 1 + c_1 z^{-1} + c_2 z^{-2} + \cdots + c_{nc} z^{-nc}\ . \end{aligned} \tag{5.9}$$

When no specific disturbance model is used, polynomial C is chosen to be 1 which is sufficient to account with the influence of white noise or a random walk disturbance in the system [6]. Considering only the disturbance part of Eq. (5.8), we can rewrite it as follows:

$$n(k) = \frac{1}{\Delta}v(k) \ \equiv n(k) = n(k-1) + v(k). \tag{5.10}$$

In fact, the integrator present in the disturbance model introduces the required integral action into the GPC control law so that one estimates the system's steady-state output and get offset-free reference tracking. This feature is more clear if we represent the model in an incremental fashion. The model, known as Controlled

Auto-Regressive Incremental Moving Average (CARIMA) model and represented in Eq. 5.8, relates the output signal with the control increments $\Delta u_k = u_k - u_{k-1}$.

$$\Delta A(z^{-1})y(k) = B(z^{-1})z^{-d}\Delta u(k-1) + C(z^{-1})v(k)\ , \tag{5.11}$$

This representation can be obtained by subtracting two consecutive estimations ($\hat{y}(k)$ and $\hat{y}(k-1)$), effectively eliminating the influence of non zero mean unknown disturbances due to the relation stated in Eq. (5.10) (assuming its average value remains constant, namely $n(k+1) = n(k)$). Consequently, the only perturbation remaining is $v(k)$ which is zero mean and its influence can be assumed null in the future time-steps.

Based on the CARIMA model previously presented, there exist several different ways of deriving the prediction equations necessary to minimize the cost function (5.2). A straightforward and transparent approach is to make use of the one-step ahead predictor:

$$\hat{y}(k+1|k) = \underbrace{-\tilde{a}_1 y(k) - \cdots - \tilde{a}_{na} y(k-na-1) + b_2\Delta u(k-1) + \cdots + b_{nb}\Delta u(k-nb)}_{\text{free response}} + \underbrace{b_1\Delta u(k)}\ , \tag{5.12}$$

where the $\tilde{a}$ coefficients are obtained as:

$$\tilde{A}(z^{-1}) = \Delta A(z^{-1})\ . \tag{5.13}$$

By recursively deriving successive predictors, a compact matrix/vector form can be written explicitly in terms of the model coefficients as:

$$\underset{k\rightarrow}{y} = \underbrace{G \underset{k\rightarrow}{\Delta u}}_{\text{forced response}} + \underbrace{F \underset{\leftarrow k}{y} + G' \underset{\leftarrow k-1}{\Delta u}}_{\text{free response}}\ . \tag{5.14}$$

Details regarding the derivation of the G, F and G' matrices (the latter is obtained from G) based on the successive recursion of the predictor can be found in [6]. While the underlying principles of this approach are more tractable for the user, the method's notation may become cumbersome for large prediction horizons. Thus, a more efficient way of coding this process for an arbitrary prediction and control window advocates the Diophantine methods [18], as will be hereby presented.

In order to obtain an explicit expression for prediction model ($\hat{y}$) for any future sampling instant p, one has to consider the Diophantine equation

$$1 = E_p(z^{-1})\tilde{A}(z^{-1}) + z^{-p}F_p(z^{-1})\ . \tag{5.15}$$

for the time instant p assuming the disturbance polynomial C equal to 1.

Following, the polynomials $E_p(z^{-1})$ and $F_p(z^{-1})$,

$$E_p(z^{-1}) = e_{p,0} + e_{p,1}z^{-1} + \cdots + e_{p,p-1}z^{-(p-1)} , \tag{5.16a}$$

$$F_p(z^{-1}) = f_{p,0} + f_{p,1}z^{-1} + \cdots + f_{p,na}z^{-na} , \tag{5.16b}$$

are of order $p - 1$ (with $(d + 1) \leq p \leq N_p$) and na respectively, and are obtained dividing 1 by $\tilde{A}(z^{-1})$ until the remainder can be factorized as $z^{-p}F_p(z^{-1})$.

Multiplying Eq. (5.11) by $E_p(z^{-1})z^p$, and considering that $v(k)$ is white noise, one obtains:

$$E_p(z^{-1})\tilde{A}(z^{-1})y(k+p) = E_p(z^{-1})B(z^{-1})\Delta u(k+p-d-1) + E_p(z^{-1})v(k+p) . \tag{5.17}$$

If we consider Eq. (5.15), then (5.17) can be rewritten as:

$$(1 - z^{-p}F_p(z^{-1}))y(k+p) = E_p(z^{-1})B(z^{-1})\Delta u(k+p-d-1) + E_p(z^{-1})v(k+p) \tag{5.18}$$

and subsequently, simplified as:

$$y(k+p) = F_p(z^{-p})y(k) + E_p(z^{-1})B(z^{-1})\Delta u(k+p-d-1) + E_p(z^{-1})v(k+p) . \tag{5.19}$$

Since the degree of the polynomial $E_p(z^{-1}) = p - 1$, the noise terms in Eq. (5.19) are all in the future. Therefore, the best prediction of $y(k + p|k)$ is given by:

$$\hat{y}(k+p|k) = G_p(z^{-1})\Delta u(k+p-d-1) + F_p(z^{-1})y(k) , \tag{5.20}$$

where $G_p(z^{-1}) = E_p(z{-}1)B(z^{-1})$.

The polynomials E_{p+1} and F_{p+1} can be obtained by the same procedure previously presented and can ultimately be derived recursively as:

$$E_{p+1}(z^{-1}) = E_p(z^{-1}) + e_{p+1,j}z^{-j} , \tag{5.21}$$

where $e_{p+1,j} = f_{p,0}$. The coefficients of the polynomial $F_{p+1}(z^{-1})$ can be recursively obtained as:

$$f_{p+1,i} = f_{p,i+1} - f_{p,0}\tilde{a}_{i+1} \qquad i = 0, \ \cdots , \ na - 1 , \tag{5.22}$$

where $f_{p,na} = 0$. The polynomial G_{p+1} can be obtained recursively as:

$$\begin{aligned} G_{p+1} &= E_{p+1}B = (E_p + f_{p,0}z^{-j})B \\ &= G_p + f_{p,0}z^{-j}B \ . \end{aligned} \tag{5.23}$$

That is, the first j coefficients of G_{p+1} will be identical to those of G_p and the remaining ones will be given by:

$$g_{p+1,p+i} = g_{p,p+i-1} + f_{p,0}b_i \qquad i = 1, \cdots, nb \ . \tag{5.24}$$

To initialize the recursion of Eq. (5.15) for $p = d+1, \cdots, N_p$ iterations, one has to consider the following initial values for the polynomials $E_{d+1}(z^{-1})$, $F_{d+1}(z^{-1})$ and $G_{d+1}(z^{-1})$:

$$E_{d+1} = 1, \tag{5.25}$$

From Eq. (5.15) it comes:

$$F_{d+1} = z^{d+1}(1 - \tilde{A}(z^{-1})), \tag{5.26}$$

and,

$$G_{d+1} = E_{d+1}(z^{-1})B(z^{-1}) = B(z^{-1}). \tag{5.27}$$

Considering now the following set of j ahead optimal predictions:

$$\begin{aligned} \hat{y}(k+d+1|k) &= G_{d+1}(z^{-1})\Delta u(k) + F_{d+1}(z^{-1})y(k) \ , \\ \hat{y}(k+d+2|k) &= G_{d+2}(z^{-1})\Delta u(k+1) + F_{d+2}(z^{-1})y(k) \ , \\ &\vdots \\ \hat{y}(k+d+N_p|k) &= G_{d+N_p}(z^{-1})\Delta u(k+N_p-1) + F_{d+N_p}(z^{-1})y(k) \ , \end{aligned} \tag{5.28}$$

they can be rewritten in a more compact form as in Eq. (5.29) (as previously presented in Eq. (5.14))

$$\mathbf{y} = \underbrace{\mathbf{G}(z^{-1})\mathbf{u}}_{\text{forced response}} + \underbrace{\mathbf{F}(z^{-1})y(k) + \mathbf{G'}(z^{-1})\Delta u(k-1)}_{\text{free response}} \ , \tag{5.29}$$

where:

$$\mathbf{y} = \begin{bmatrix} \hat{y}(k+d+1) \\ \hat{y}(k+d+2) \\ \vdots \\ \hat{y}(k+d+N_p) \end{bmatrix}, \quad \mathbf{u} = \begin{bmatrix} \Delta u(k) \\ \Delta u(k+1) \\ \vdots \\ \Delta u(k+N_u-1) \end{bmatrix},$$

$$\mathbf{G}(z^{-1}) = \begin{bmatrix} g_{1,0} & 0 & \cdots & 0 \\ g_{2,1} & g_{2,0} & \cdots & 0 \\ g_{3,2} & g_{3,1} & \ddots & \vdots \\ g_{N_p,N_p-1} & g_{N_p,N_p-2} & \cdots & g_{N_p,N_p-N_u} \end{bmatrix}, \quad \mathbf{F}(z^{-1}) = \begin{bmatrix} F_{d+1}(z^{-1}) \\ F_{d+2}(z^{-1}) \\ \vdots \\ F_{d+N_p}(z^{-1}) \end{bmatrix},$$

$$\mathbf{G}'(z^{-1}) = \begin{bmatrix} (G_{d+1}(z^{-1}))z \\ (G_{d+2}(z^{-1}) - g_0 - g_1 z^{-1})z^2 \\ \vdots \\ (G_{d+N_p}(z^{-1}) - g_0 - g_1 z^{-1} - \cdots - g_{N_p-1} z^{-(N_p-1)})z^{N_p} \end{bmatrix}.$$

Based on the cost minimization procedure generically presented in Sect. 5.2, the optimal control increments are obtained by:

$$\mathbf{u} = (\mathbf{G}^T\mathbf{G} + \lambda\mathbf{I})^{-1}\mathbf{G}^T(\mathbf{R} - \mathbf{F}y(k) - \mathbf{G'}\Delta u(k-1)) , \tag{5.30}$$

where $\mathbf{R} = [r(k+d+1), \cdots, r(k+N_p))]$ is the reference set-point in the N_p prediction horizon.

As GPC is implemented in a receding horizon fashion, the control signal sent to the process is incremented by the first value of **u**, $\Delta u(k)$, which is given by:

$$\Delta u(k) = \mathbf{K}(\mathbf{R} - \mathbf{F}y(k) - \mathbf{G'}\Delta u(k-1)) , \tag{5.31}$$

where **K** is the first row of matrix $(\mathbf{G}^T\mathbf{G} + \lambda\mathbf{I})^{-1}\mathbf{G}^T$,

$$K = [1\ 0\ 0\ \cdots 0]_{N_u}(\mathbf{G}^T\mathbf{G} + \lambda\mathbf{I})^{-1}\mathbf{G}^T . \tag{5.32}$$

Following this approach, the accuracy problems that the prediction model may present when extrapolating the system's response in long time horizons are minimized. This is due to the constant corrections of the optimal control trajectory performed according to the most recent system samples.

5.4 Extension of Generalized Predictive Control to Non-linear Models

The predictor derived in the previous section assumes the existence of a linear plant model complete enough to fully capture the process dynamics. If the process non-linearity is not significant, then the integral action of the linear GPC algorithm will compensate eventual inaccuracies and eliminate the steady-state error. However, since some physical processes possibly present complex non-linear relations, a simple linear model may prove itself insufficient for a successful application of the GPC algorithm. The mismatch between the model and the corresponding physical process is more significant in situations where operation point of the process changes fast and significantly, resulting in slow or unstable closed loop response in such operation regions [3]. For this reason, to apply the GPC principles in domains where its success has been hindered by the non-linear behavior of the process, several extensions to its original linear formulation have been proposed in recent years. Though, it is important that such enhancements, apart from improving the performance of the closed loop system, maintain the similar level of simplicity and effectiveness that traditional linear GPC accustomed its practitioners in industry.

The most straightforward way of dealing with GPC in non-linear processes is to directly use a single non-linear model describing every operational regime and to solve an optimization problem that yields the control trajectory that minimizes a considered cost function. Models based on ANNs are particularly successful for such purpose due to their regular structure and universal approximation capabilities [3]. However, without any further simplifications of the model, a non-linear GPC optimization problem has to be solved on-line at each sampling instant. Apart from being computationally demanding, non-linear optimization algorithms neither guarantee that a global optimal solution is found due to the non-convex nature of the problem (which may result in unsatisfactory control quality), nor that convergence to a sub-optimal one is obtained in useful time (what may hinder their use in on-line GPC implementations requiring short sampling periods).

Despite the increasingly availability of powerful embedded systems in the market, practical implementations of GPC for non-linear processes circumvent this problem by relying on linear approximations of the non-linear model. This approach, known as instantaneous linearization [3], computes at each sampling instant a linear model of the process so that one can obtain a estimation of the free and forced responses of the process over the prediction horizon. By using this procedure it is possible to derive an explicit control law using a similar process as previously presented on Sect. 5.3. The idea of such approach is presented on Fig. 5.5.

As is clear from the difference in the amount of publications between both lines of work, "theoretical purists tend to stay away from linearization approaches" [3].

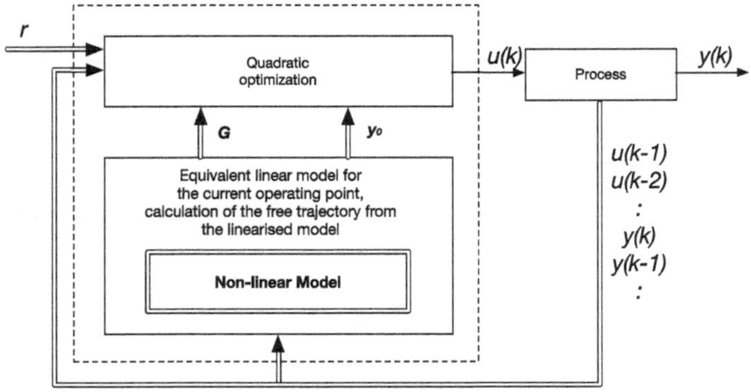

Fig. 5.5 Structure of the GPC algorithm based on a linearization of a non-linear model

The main reason for such conduct is that linearization procedures make the stability and robustness analysis much more difficult than in the case of GPC based on pure non-linear models due to the possibility of existing discontinuities in the parameters of the models underlying the controller synthesis procedure [10].

There are two main lines of work regarding the use of linearized models in GPC. In the first approach, one calculates a linear approximation of the non-linear model for the current operating point and consider its parameters constant for the whole prediction horizon. Yet, in some scenarios where changes in the operation region are significant, there may exist a larger discrepancy between the real process trajectory and the estimated one in the end of the prediction horizon, what may lead to suboptimal results. When such assumption is not sufficient, one can alternatively perform the linearization along a future input trajectory to account with the possible model variations. Since such trajectory is not known a priori, the optimization process requires a larger number of iterations within the prediction horizon to converge towards the optimal control increments. Nevertheless, when the inherent limitations of GPC designs based on single linear approximation are accounted for, very little differences can be seen between the optimal trajectories obtained by both approaches [3].

As was presented in Chap. 4, TS FS proved their ability to accurately approximate any non-linear dependency between input and output variables. In the present application scenario, this modeling approach is particularly advantageous since its structure inherently considers a set of locally linearized models valid within a certain operation region defined by its non-linear fuzzy boundaries. By employing the available fuzzy inference mechanisms, one can also efficiently interpolate the output of the several local models to obtain an approximate linear model valid for the current operation regime. Performing this operation at each sampling instant, a TS Fuzzy Model can in fact be considered a linear time-varying model. For that reason, it can be easily integrated with traditional model-based control methodologies such as GPC to overcome the majority of the limitations and difficulties imposed other approaches based on non-linear models. Successful applications resulting from the synergy between Fuzzy Models and Predictive Control can already be found in literature. They span theoretical analysis such as in [10, 19], focused in the analysis of the robust stability conditions and supported with numerical simulations, up to practical ones evincing the robustness of the method to disturbances in industrial domains such as in the control of a non-linear heat-exchanger pilot plant [20], a stirred tank reactor [8] or a binary distillation column [21].

5.4.1 Generalized Predictive Control Using Type-2 TS Fuzzy Models

The extension of the GPC based on TS Fuzzy Systems to the Type-2 FL case is a straightforward procedure [22]. As was presented in Chap. 4, a Type-2 Fuzzy Model can ultimately be represented by two average sub-models related with the upper and

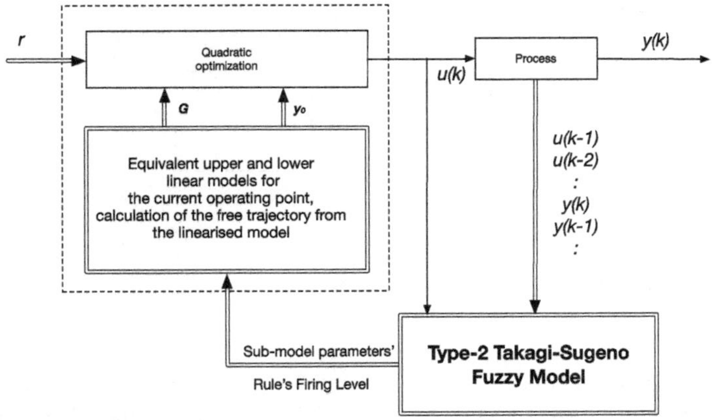

Fig. 5.6 Structure of the GPC algorithm based on Type-2 TS Fuzzy models

lower bounds of the output value uncertainty interval. Based on each sub-model, the GPC theory can be directly applied thus, two control increments will be obtained, namely $\Delta\overline{u}$ and $\Delta\underline{u}$. The effective control action performed by the controller is then given by averaging the control increments obtained as:

$$u(k) = u(k-1) + \frac{\Delta\overline{u} + \Delta\underline{u}(k)}{2}\ . \tag{5.33}$$

In Fig. 5.6, the functional diagram of the GPC based on Type-2 TS Fuzzy Models is presented.

5.5 Application Scenarios

To evaluate the enhancements obtained by combining the modeling capabilities of IT2FLSs and the GPC theory, the Fermentation Reactor and Coupled Tanks Systems presented in the previous chapter will be used as benchmark for the proposed model based predictive controller. Additionally, its performance will be compared with two additional implementations based on an linear ARX and a Type-1 TS Fuzzy models. Figure 5.7 displays the evaluated closed loop system along with the considered disturbances affecting its operation.

The models used as support for the GPC algorithm are considered fixed during closed loop operation. While model adaptability is a highly sought feature in control systems [13], it also presents several problems due to its unpredictable influence in the model parameters and, consequently, in the closed loop behavior. Considering the particular scenarios where the closed loop control system must perform quick changes between different operation regimes, the use of adaptive models may eventually result in suboptimal or even unstable control as the model update rate

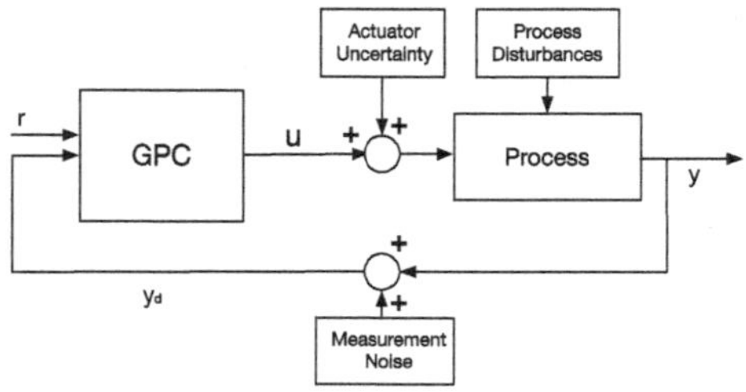

Fig. 5.7 Diagram of the closed loop control system

may be insufficient to cope with the significant changes verified on some parameters. Nevertheless, by considering the compared models fixed, the advantages of the GPC control algorithm based on Type-2 TS Fuzzy models in coping with inherent modeling uncertainties will be better highlighted. In the evaluations performed, the dynamic equations of the benchmark systems were implemented in continuous time while the GPC is executed in discrete time.

5.5.1 *Fermentation Reactor's Temperature Control*

As was previously presented, from the perspective of a control algorithm the Fermentation Reactor is a single-input single-output process, the coolant flow rate (F_{ag}) the manipulated variable and the reactor's temperature (T_r) the controlled one, as shown in Fig. 5.8.

The maximum allowed cooling fluid's flow rate is 80 l h^{-1} and the reactor's temperature is controlled within the [28–33] °C interval. The performance of the GPC in the Fermentation Reactor's Temperature Control will be assessed considering the following operation scenarios:

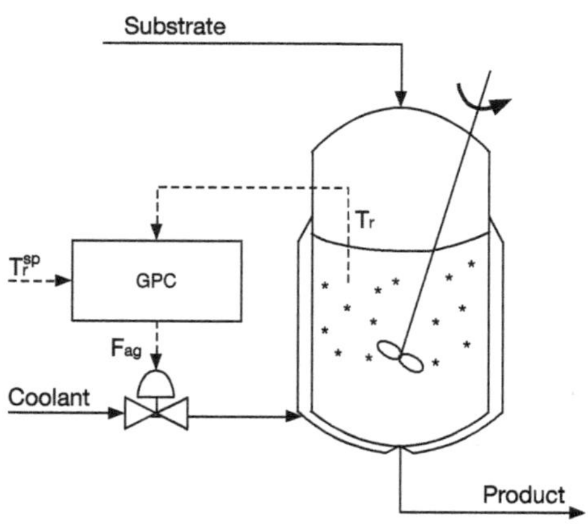

Fig. 5.8 Diagram of the Fermentation Reactor

- Different amplitude reference step signals are used by closed loop system as control goals. An unmeasured gaussian disturbance with zero mean and standard deviation of $0.1\,\mathrm{l\,h^{-1}}$ is added to the control signal while the reactor's temperature measurements are corrupted by a gaussian disturbance with zero mean and $0.05\,°\mathrm{C}$ standard deviation. The system's model is trained under these conditions.
- Under similar operation conditions of the previous scenario, a step change in the substrate temperature from 25 to 27 °C is introduced. This type of disturbance can occur due to ambient temperature changes.

In the evaluated controllers, the considered prediction and control horizons are 10 and 3 samples respectively and the control signal updated once every hour. When employing TS fuzzy models the control activity penalty factor (λ_{GPC}) is set to 0.01. Its value is changed to 0.1 when the ARX model is adopted. These parameters are chosen so that a control signal with satisfactory transient response is attained while maintaining its robustness to noise. The difference in the control activity penalty factor between the ARX and the TS Fuzzy based implementations already anticipates the superiority of the latter approaches since they allow one to synthesize a faster control law for identical disturbance levels. As was presented in the previous chapter, the TS structures used to implement the controllers are composed by 5 rules. The Type-2 TS Model used considered 5 and 12% uncertainty factors over the antecedent and consequent part parameters, respectively. The obtained results are evaluated using four distinct metrics: the Mean Squared Error (MSE), the overshoot (O_S), the settling time (T_S) of the controlled variable and the Control Effort (CE) of the actuation variable. The latter one is estimated by the cost function as:

$$CE = \sum_{k=t}^{t+N} \frac{\Delta u(k)^2}{N} \tag{5.34}$$

In Fig. 5.9, the closed loop response of the system is presented considering different set-points and operating conditions identical to the training stage.

This evaluation scenario shows that the closed loop responses of the ts Fuzzy Model based controllers present a similar transient response, standing out a slight improvement when the Type-2 one is used. Despite the worse performance of the linear arx based controller during the step transient stage, it is also capable of achieve zero steady-state error. This result attests the robustness of GPC even in the case of significant model mismatches (as long they are compensated with an adequate control activity penalty factor). A detailed view of the system's behavior when the set-point is changed from 29°C to 31.5°C is presented in Fig. 5.9, and the comparative metrics relative to the three control systems are presented in Table 5.1.

In this comparison scenario, the Type-2 TS based GPC achieves improved metrics relatively to the Type-1 and ARX based controllers, reducing the MSE over the evaluated interval by approximately 6% and 65%, respectively. Additionally, there is a significant improvement in the step response overshoot and settling time which, in this specific application, leads to a reduction of the transient response length by several hours. As there is a large dependence of the reaction sub-products' parameters on

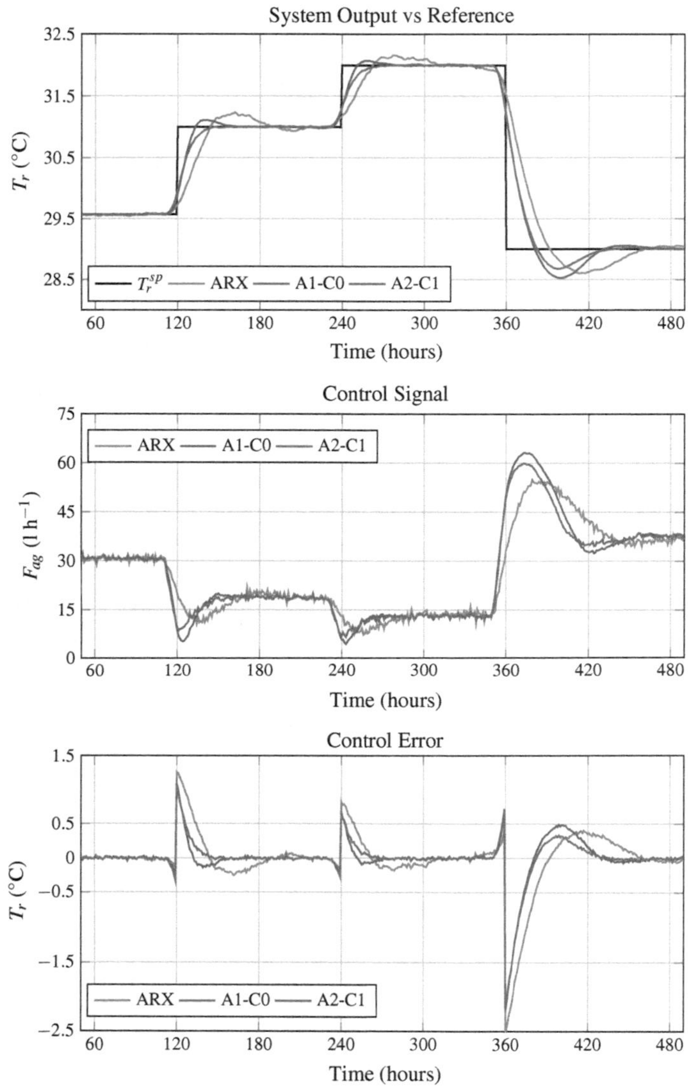

Fig. 5.9 Closed loop behavior of the reactor's temperature during a yeast fermentation reaction using three different GPC algorithms based on locally linear models

the reactor's temperature, the use of a faster controller allows one to perform a quick change between different operation regimes as production requirements dictate.

To evaluate the disturbance rejection capabilities of each controller and its performance under model mismatches, several set-point changes were performed after the raw material's temperature at the reactor's intake is increased from 25 to 27 °C at instant $t = 50$ h, as shown in Fig. 5.11. The obtained results reveal that the Type-2

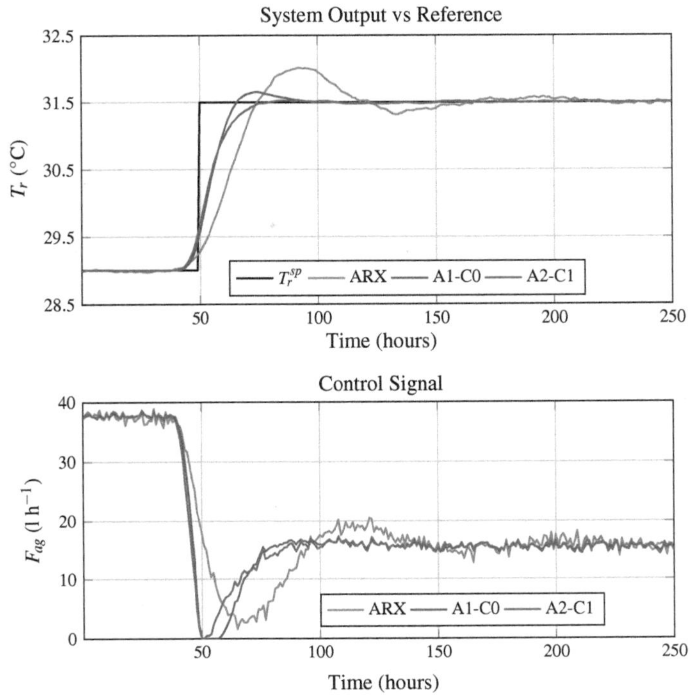

Fig. 5.10 Detailed view of the process's response to a change in the controller's reference signal

Table 5.1 Comparative metrics under nominal operation measured at reference step (29.0 to 31.5 °C) as presented in Fig. 5.10

	MSE ($\times 10^{-2}$)	CE	O_s (°C)	T_s (h)
ARX	19.50	10.2	0.499	180
A1-C0	7.54	14.6	0.153	53
A2-C1	6.81	11.9	0.028	48

TS GPC controller provides a faster transient response when external perturbations interfere with the reactor's temperature. Additionally, its advantages are more pronounced when the controller is operating in a different regime that the one it was trained for. Despite the significant change on the cooling fluid's flow rate required during the period [0–50] h, the Type-2 TS based controller provides a more "desirable" closed loop response. Table 5.2 briefly summarized the comparative metrics obtained for the three controllers.

Ultimately, to assess the model dimensionality reduction capabilities of Type-2 Fuzzy Models, the 5 Rule Type-2 TS Fuzzy Model was compared with a 9 Rule Type-1 TS one. By increasing the number of rules of the latter model, one expects it

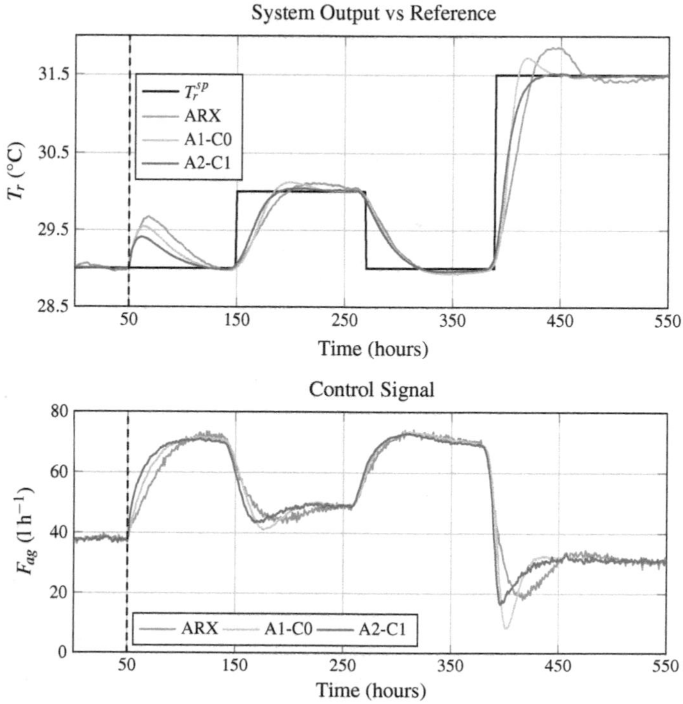

Fig. 5.11 Evaluation of the closed loop performance after the process is disturbed by a change in the substrate temperature

Table 5.2 Comparative metrics measured at reference step (29 to 31.5 °C) after introduction of a disturbance, as presented in Fig. 5.11 in the interval [350–550] h

	MSE ($\times 10^{-1}$)	CE	O_s (°C)	T_s (h)
ARX	4.15	14.6	0.350	130
A1-C0	3.14	31.4	0.275	100
A2-C1	2.56	30.4	0.051	68

to perform similarly to its Type-2 counterpart. Figure 5.12 displays their performance under nominal operating conditions.

Establishing a comparative standpoint with the previous evaluations, Fig. 5.13 provides a depiction of the system's response when the reference signal changes from 29 to 31.5 °C. The comparative metrics are summarized in Table 5.3.

The results demonstrate that the controller based on the Type-1 TS Fuzzy model achieves a transient response closer to the one obtained with the Type-2 model in the ascending transitions. This improvement comes at the cost of an slight increase of control signal activity, what is not necessarily good as the system's noise robustness may be reduced. It is observed though that the descending transitions of the system's

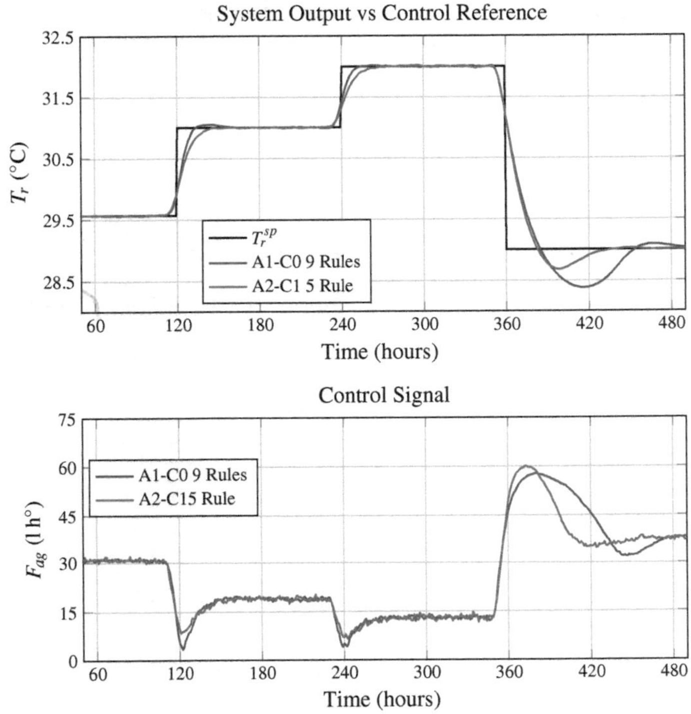

Fig. 5.12 Comparison between two predictive controllers based on a 5 Rule Type-2 TS Fuzzy model and a 9 Rule Type-1 TS Fuzzy model ($N_p = 10$; $N_u = 3$, $\lambda_{GPC} = 0.01$)

Table 5.3 Comparative metrics under nominal operation measured at reference step (29.0 to 31.5 °C) presented in Fig. 5.13

	MSE ($\times 10^{-2}$)	CE	O_s (°C)	T_s (h)
A1-C0 (5 Rules)	7.54	14.6	0.153	53
A1-C0 (9 Rules)	8.64	37.4	0.051	49
A2-C1 (5 Rules)	6.81	11.9	0.028	48

response is slightly degraded comparing to the controller based on the 5 Rule Type-1 TS Fuzzy model (presented in Fig. 5.9). This behavior is due to the non-linear nature of the Reactor's temperature model in the warming-up/cooling-down stages.

Evaluating now the system's behavior after increasing the raw material's temperature from 25 to 27 °C, the differences between both modeling approaches become more clear as depicted in Fig. 5.14. The comparative metrics are presented in Table 5.4.

With these results, one may conclude that the Type-2 TS Fuzzy Model improvements are not solely related to its equivalence to a larger Type-1 Fuzzy Model. The Type-Reduction mechanism establishes a dependency between the uncertainty

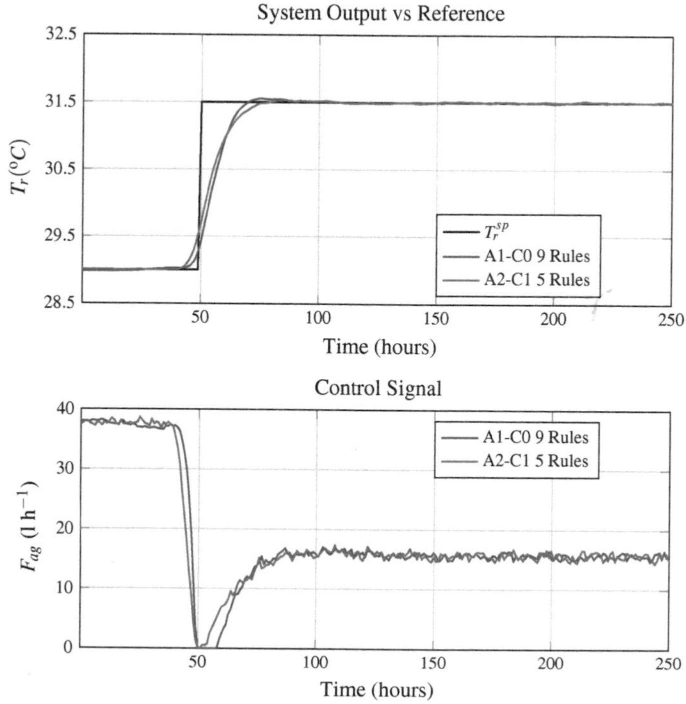

Fig. 5.13 Detailed view of the process's closed loop response under identical conditions to the training procedure using a GPC based on a 9 Rule Type-1 and a 5 Rule Type-2 Fuzzy model

Table 5.4 Comparative metrics under constant disturbance: measured at reference step (29.0–31.5 °C) after introduction of a disturbance, as presented in Fig. 5.14 in the interval [350–550] h

	MSE ($\times 10^{-1}$)	CE	O_s °C)	T_s (h)
A1-C0 (5 Rules)	3.14	31.4	0.275	102
A1-C0 (9 Rules)	6.73	10.1	0.122	112
A2-C1 (5 Rules)	2.56	30.4	0.051	68

degrees of the input space partition (defined by the upper and lower bounds of a rule's firing level) and the system's output that has more degrees of freedom than those obtained with a Type-1 TS Fuzzy Model, ultimately leading to systems which perform differently. Therefore, it is not expected to attain a proportional relationship between the number of rules and the performance metrics obtained of the two types of TS Fuzzy Models.

The manipulation of the additional degrees of freedom provided by Type-2 FL naturally requires a superior computational complexity comparatively to its Type-1 counterpart. Therefore, Table 5.5 presents an overview of the mean control loop execution time and its standard deviation for the two modeling approaches. The obtained

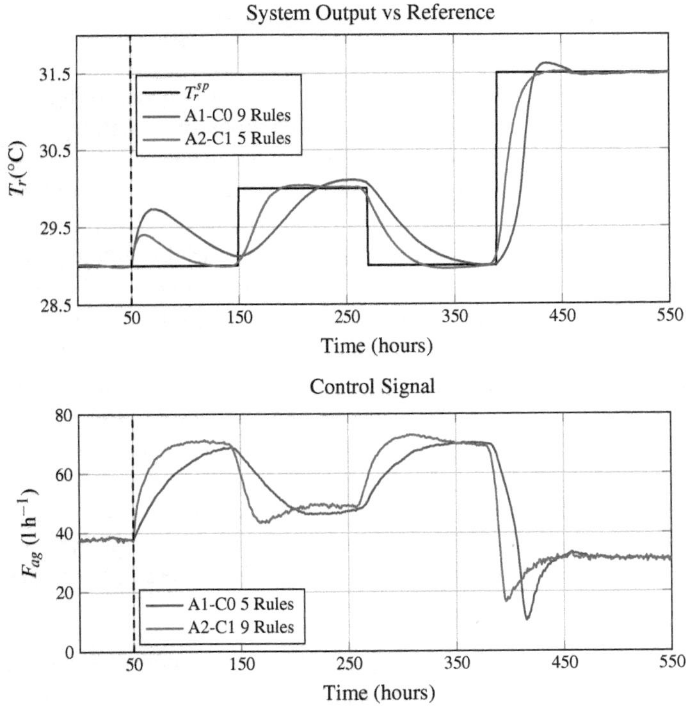

Fig. 5.14 Comparison between two predictive controllers based on a 5 Rule Type-2 TS Fuzzy model and a 9 Rule Type-1 TS Fuzzy model ($N_p = 10$, $N_u = 3$, $\lambda_{GPC} = 0.01$) after the raw material's temperature (T_{in}) is changed from 25 to 27 °C at instant $t = 50$ h

Table 5.5 Execution time of the GPC algorithm based on TS Fuzzy models

	5 rule	5 rule	9 rule
	A1-C0 model	A2-C1 model	A1-C0 model
Mean	8.74×10^{-4}s	15.35×10^{-4}s	10.22×10^{-4}s
Std. Deviation	5.08×10^{-9}s	1.78×10^{-8}s	6.86×10^{-9}s

metrics refer to the execution of the one step-ahead predictor and the synthesis of the control law, resulting from the average of 56 set-point changes.

The use of a GPC based on the 5 Rule Type-2 TS Fuzzy Models requires a larger computational effort, approximately 1.76 and 1.5 times larger than the 5 Rule and 9 Rule Type-1 TS Fuzzy models, respectively. Nevertheless, after the analysis performed in this chapter, one may conclude that the Type-2 TS Model based controller presents in this scenario an improved servo performance and disturbance rejection that, when no computational time constraints are present, makes it the preferred approach.

5.5.2 Coupled Tanks Liquid Level Control

The Coupled Tanks System is a setup widely used for benchmarking control algorithms [23, 24] due to the multitude of combinations of actuation and controlled variables that can be defined upon. As was presented in Chap. 4, by choosing which pumps and rotary valves are directly manipulated, one can develop either a SISO or a MIMO control system to control the liquid level of one of the tanks. In the scenario presently evaluated, the Pump 1 will be the actuation variable and the liquid level of the Tank 2 the controlled variable. The maximum flow rate for the Pump 1 is 90 cm^3 s^{-1}, while the Pump 2 (which will act as an unmeasured disturbance) is limited to 20 cm^3 s^{-1} and the liquid height in Tank 2 controllable in the [0,45]cm interval. By coupling solely two reservoirs, this setup can be seen as a non-linear second order SISO system, where the remaining actuators available are unmodeled disturbances which will be used to evaluate the robustness of the developed control algorithm. Figure 5.15 depicts the setup used in the present test.

Similarly to the application scenario previously evaluated, three predictive controllers' implementations based on linear ARX, Type-1 TS and Type-2 TS Fuzzy Models will be compared. The results further presented were obtained according to the following test scenarios:

- Different amplitude reference step signals are used. An unmeasured gaussian disturbance is introduced in the control signal of Pump 1 with zero mean and variance of 0.5 cm^3 s^{-1} while the Tank 2 liquid level measurements are corrupted by a gaussian noise with zero mean and variance of 0.05 cm. The Pump 2 is turned off during this evaluation. The system's model is trained under these operation conditions.

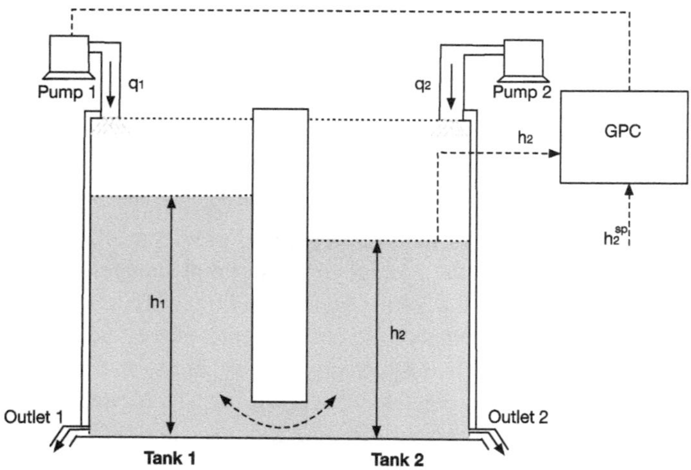

Fig. 5.15 Diagram of the coupled tanks control system

Table 5.6 Comparative metrics under nominal operation measured during the reference step (15 to 25 cm) presented in Fig. 5.16

	MSE	CE	O_s (cm)	T_s (s)
ARX	3.20	32.7	2.5	180
A1-C0	4.55	6.53	5.1	302
A2-C1	2.48	4.96	0.4	102

- Under similar conditions of the previous test, the Pump 2 is used to introduce a gain change in the closed loop system by turning it on with a constant flow rate of $20\,\text{cm}^3\,\text{s}^{-1}$, directly disturbing the Tank 2 liquid level.
- Ultimately, maintaining the actuation and measurement disturbance levels of the initial test, the Pump 2 is used to introduce a low frequency disturbance in the closed loop system. To do so, Pump 2 is controlled so its flow rate varies sinusoidally in the interval [0, 16] $\text{cm}^3\,\text{s}^{-1}$ with a period of 800 s.

In the three GPC implementations, the prediction and control horizons were 10 and 3 samples and the control activity penalty factor $\lambda_{GPC} = 0.5$. The controller algorithm updates its output every 2 s (at the same rate of the sampling interval). According to the models developed in the previous chapter, a good compromise between modeling accuracy and dimensionality is achieved using TS Fuzzy Models with 4 rules. The Type-2 TS Model used considered 5% and 8% uncertainty factors over the antecedent and consequent part parameters, respectively.

Regarding the first evaluative scenario, the response of the plant to the several control systems is presented in Fig. 5.16. The advantages of the GPC controller based on the Type-2 TS Fuzzy Model comparing to the remaining ones are clear, providing a closed loop response with the fastest settling time and minimal overshoot (close to a critically damped behavior) without saturating the control signal. Table 5.6 overviews the evaluated metrics, obtained during the [1300–1900] s time interval. When comparing the controllers based on the linear ARX and the Type-1 TS models, at first instance the former approach seems more advantageous. However, such results come at a cost of a significantly higher control effort given the linear ARX model mismatches (highlighted in the Chap. 4). As so, to attain a more stable control signal, this penalty factor would have to be increased ultimately yielding a slower transient response.

To test the model based controllers when the plant deviates from the nominal operation conditions, the Pump 2 is manipulated to supply a constant flow rate of $20\,\text{cm}^3\,\text{s}^{-1}$. As depicted in Fig. 5.17, this disturbance is activated at the time instant $t = 200$ s, introducing a change in the plant's steady state gain.

Similarly to the previous evaluation, the GPC based on the Type-2 TS Fuzzy Model maintains a superior closed loop performance, presenting faster disturbance suppression and a critically damped behavior on the subsequent set-point transitions. Table 5.7 presents an overview of the obtained metrics.

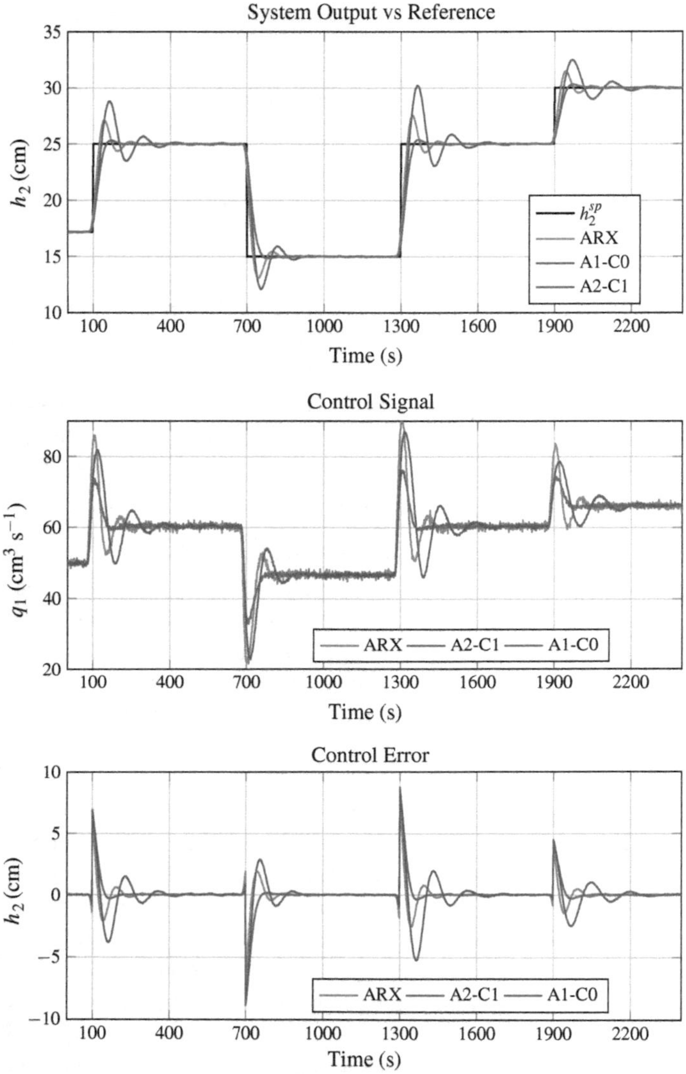

Fig. 5.16 Closed loop behavior of the coupled tanks system using three different GPC algorithms based on locally linear models

Finally, the process is subjected to an unmeasured sinusoidal disturbance introduced by the Pump 2. Its flow rate was manipulated sinusoidally, with an amplitude varying within the [0–16] $\mathrm{cm}^3\ \mathrm{s}^{-1}$ interval and a period of 800 s. This low frequency disturbance was considered to be in the region of interest of the system, considering the settling time values during nominal operation conditions. As depicted in Fig. 5.18, the sinusoidal disturbance was significantly attenuated in every controller implementation, existing a maximum ripple of approximately 0.35 cm around the

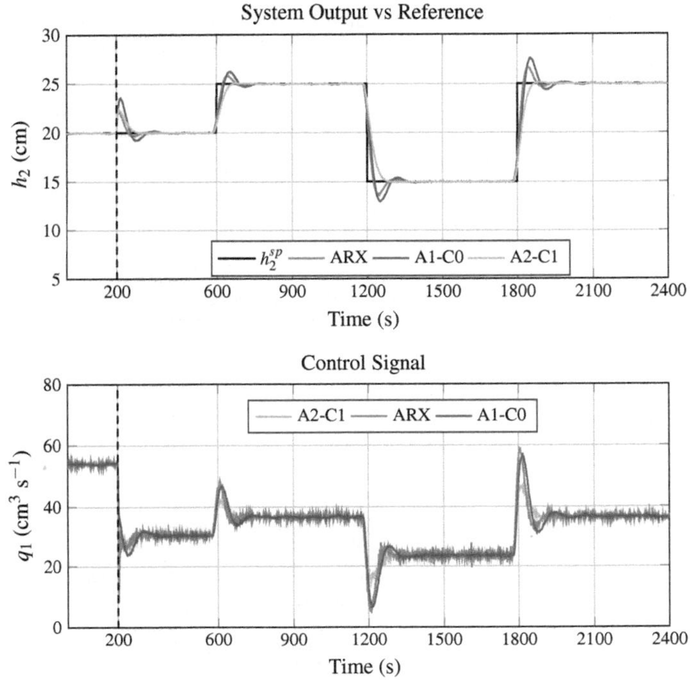

Fig. 5.17 Evaluation of the closed loop performance after the process is disturbed by Pump 2 with a constant liquid flow rate ($h_2 = 20$ cm^3 s^{-1}) after $t = 200$ s

Table 5.7 Comparative metrics under constant disturbance: measured at reference step (15 to 25 cm) presented in Fig. 5.17

	MSE	CE	O_s (cm)	T_s (s)
ARX	3.32	33.4	1.7	162
A1-C0	3.79	3.79	2.6	311
A2-C1	2.05	3.37	0.2	139

reference signal. Yet, the TS Fuzzy Models based ones perform significantly better in terms of the required control effort. The comparative metrics presented in Table 5.8.

The evaluation scenarios hereby presented demonstrated the importance of the control activity penalty factor on the control signal's robustness. While at first instance the closed loop system output behavior based on the linear ARX system is very close to the Type-1 TS based one, such result comes at the expense of an increased actuator's activity cost. A coarser model may provide sufficient information about the system's behavior trend, hence a well dampened control law is capable of control the system's output. However, for a finer control quality and faster transient

Table 5.8 Comparative metrics under sinusoidal disturbance: MSE, control effort (CE), overshoot (O_S) and oscillation amplitude at the output (O_a) measured after the reference step (15–25 cm) presented in Fig. 5.18

	MSE	CE	O_s (cm)	O_a (cm)
ARX	2.70	27.35	1.82	0.15
A1-C0	2.16	7.16	2.98	0.35
A2-C1	2.08	4.23	0.02	0.3

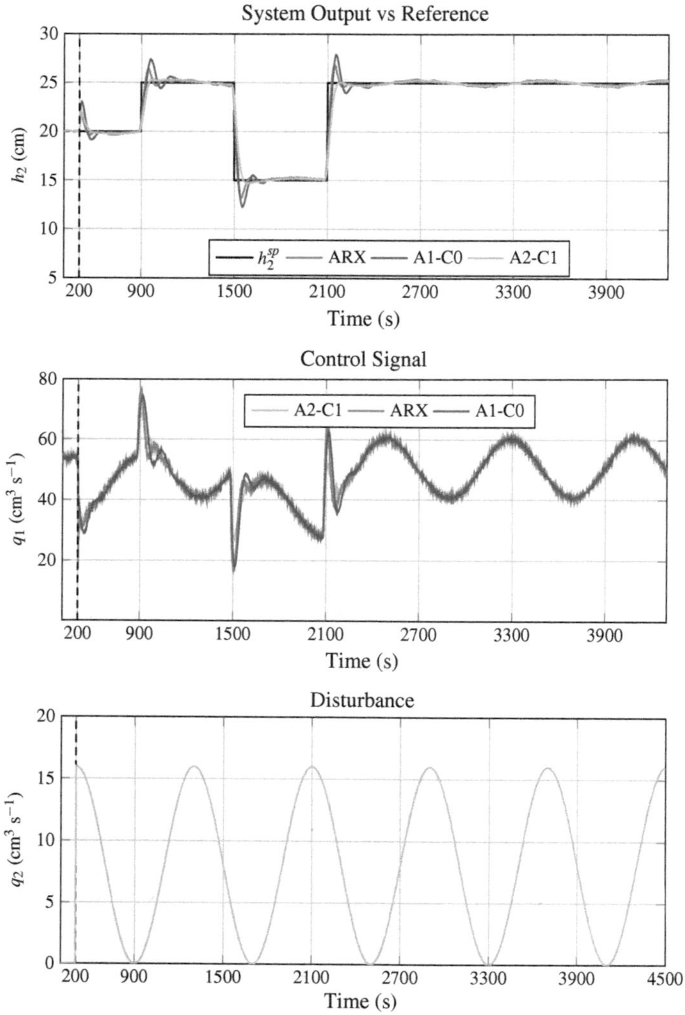

Fig. 5.18 Evaluation of the closed loop performance after the process is disturbed by Pump 2 with a sinusoidal flow rate (q_2)

response, a more accurate model is required. Therefore, the results attained justify the choice of the approach based on Type-2 TS Fuzzy models.

5.6 Conclusions

The procedure hereby described to extend the use of a Type-2 Fuzzy Model to Model Predictive Controllers can be interpreted as an instantaneous linearization of the process dynamics on the current operating point. The process non-linearity is naturally embedded on the firing level that weights the contribution of each sub-model to the final one, and ultimately leads to the development of a simple linear structure with variable parameters. As the modeled plant is of non-linear nature, the validity of the global model predictions is restricted to a limited operation region. Hence, GPC algorithms based on linearized models are inherently sub-optimal because their predictions are likely to be different from those obtained by the original non-linear one. Thus, one must not rely too heavily on the linearized model (as when using long prediction horizons) and design a controller that does not violate the limitations of the approximation such as avoiding abrupt changes in the operation region by simply limiting the process's set-point slew-rate. When such limitations are considered, one has a computationally efficient non-linear GPC framework capable of achieve well performing closed loop control systems.

References

1. Maciejowski, J.: Predictive Control with Constraints. Prentice Hall (2001)
2. Quin, S., Badgwell, T.: A survey of industrial model predictive control technology. Control Eng. Pract. **11**(7), 733–764 (2003)
3. Ławryńczuk, M.: Computationally Efficient Model Predictive Control Algorithms - A Neural Network Approach, Studies in Systems, Decision and Control, vol. 3. Springer, Heidelberg (2014)
4. Del Re, L., et al.: Automotive Model Predictive Control - Models. Lecture Notes in Control and Information Sciences, Methods and Applications. Springer, Verlag (2010)
5. Hafez, A.: Design and Implementation of Modern Control Algorithms for Unmanned Aerial Vehicles, Ph.D. Thesis, Queen's University, Ontario, Canada (2014)
6. Rossiter, J.: Model-Based Predictive Control: A Practical Approach (Control series). CRC Press, Boca Raton (2004)
7. Mollov, S., Babuška, R., Abonyi, J., Verbruggen, H.: Effective optimization for fuzzy model predictive control. IEEE Trans. Fuzzy Syst. **12**(5), 661–675 (2004)
8. Mendes, J.; Araujo, R.; Souza, F.: Adaptive Fuzzy Identification and Predictive Control for Industrial Processes, Expert Systems with Applications, vol. 40, pp. 6964–6975, Elsevier (2013)
9. Sousa, J., Kaymak, U.: Model prediction control using fuzzy decision functions. IEEE Trans. Syst. Man Cybern. **31**(1), 54–65 (2001)
10. Mollov, S., et al.: Robust stability constraints for fuzzy model predictive control. IEEE Trans. Fuzzy Syst. **10**(1), 50–64 (2002)

11. Rouhani, R., Mehra, K.: Model algorithmic control (MAC): Basic theoretical properties. Automatica **18**, 401–441 (1982)
12. Cutler, C.; Ramaker, L.: Dynamic matrix control - a computer control algorithm. In: Proceedings of the AICHE National Meeting (1979)
13. Åström, K.; Wittenmark, B.: Adaptive Control. Dover (2008)
14. Levenberg, K.: A method for the solution of certain problems in least squares. Q. Appl. Math. **2**, 164–168 (1944)
15. Marquardt, D.: An algorithm for least-squares estimation of non-linear parameters. J. Soc. Ind. Appl. Math. **11**(2), 431–441 (1963)
16. Nocedal, J., Wright, S.: Numerical Optimization, 2nd edn. Springer, New York (2006)
17. Clarke, D., Mohtadi, C., Tuffs, P.: Generalized predictive control, parts 1 and 2. Automatica **23**, 137–160 (1987)
18. Camacho, E.; Bordons, C.: Model Predictive Control, Advanced Textbooks in Control and Signal Processing, 2nd edn. Springer, Heidelberg (2007)
19. Lu, C.: Wavelet fuzzy neural network for identification and predictive control of dynamic systems. IEEE Trans. Ind. Electron. **58**(7), 3046–3058 (2011)
20. Skrjanc, I., Matko, D.: Predictive functional control based on fuzzy model predictive control. IEEE Trans. Fuzzy Syst. **10**(1), 50–64 (2002)
21. Mahfouf, M.; Linkens, D.; Abbo, M.: Adaptive Fuzzy TSK Model-Based Predictive Control Using a CARIMA Model Structure, Chemical Engineering Research and Design - Process Control, vol. 78, no. 4, pp. 590-596. Elsevier (2000)
22. Antão, R., Mota, A., Martins, R.: Model-based control using interval type-2 fuzzy logic systems. Soft Comput. (2016)
23. Wu, D., Tan, W.: A simplified type-2 fuzzy logic controller for real-time control. ISA Trans. **45**(4), 503–516 (2006)
24. Ogata, K.: Modern Control Engineering, 5th edn. Prentice Hall (2009)

Chapter 6
Processor-In-the-Loop Simulation

6.1 Introduction

Empowered by the increasing computational power broadly available in current computer technology, the use of simulation software is an ubiquitous approach both in academia and industry during the development path of a large number of systems. Process's modeling and control is one particular domain that greatly benefited from the availability of such tool, overcoming two important constraints that dictate the course of actions taken during a new product's implementation—Time and Cost [1]. Such factors are typically related with the following problems:

- Availability of the plant
- Cost and time of building a control system prior to testing
- Difficulty in obtaining repeatable conditions during the development stage
- Time consuming testing for system's validation
- High cost due to failure

Simulation tools are also an important asset during the initial stages of control systems' development because they allow the execution of several benchmarking trial cases which can then be taken as reference for the subsequent validation stages of the algorithm implementation. However, achieving optimal results under a simulated environment is solely a partial achievement towards its deployment under real operation conditions. While the computational power of a general purpose computer is of great advantage for the simulation stages, the very same systems are not adequate for the final implementation of application specific control loops. For such purposes, embedded computer systems are better suited since their "simpler" hardware and software architectures allow low-level access to peripheral devices interfacing with actuation and sensor systems and are better compliant with the strict timing requirements that control algorithms typically demand.

Ideally, an embedded control system is tested against the real plant, but it is common to find scenarios involving several limitations and risks in the scope of the testing (such as going beyond the range of the control system parameters or

R. Antão, *Type-2 Fuzzy Logic*, Nonlinear Physical Science,
DOI 10.1007/978-981-10-4633-9_6

plant capabilities). Hence, the validation of complex algorithms implemented in such platforms presents additional engineering challenges: How can a meaningful set of test vectors that approximate the input/output behavior of a process under control be generated? How can the control algorithm behavior be analyzed in real-time under specific operation conditions? How will the algorithm's turnaround time affect the operation of the actual system? The advantages of physical process's simulation previously enumerated can be used as well to overcome such questions and, together with the algorithm's execution in an embedded platform, provide a superior test platform closer to the conditions experienced when a system is deployed in the real environment.

Hardware-In-the-Loop (HIL) simulation is a technique for performing system-level testing of embedded systems in a comprehensive, cost effective and repeatable way using a combination of electronic hardware and custom software. Such approach is mostly used in the validation of embedded systems when they cannot be tested easily, thoroughly, and repeatably in their operational environments [1]. To accomplish so, HIL replaces the plant under control with a powerful computer system capable of simulate several interconnected processes in Real-Time based on their dynamic models. In the tested system's point of view, there will be as little differences as possible comparatively to a real scenario. Figure 6.1 depicts this closed loop architecture.

The interconnection between the simulation model and the controller hardware can be implemented using A/D and D/A conversion stages based not only in analog signals that approximate the plant's sensors and the actuation systems, but also supported by common serial communication interfaces such as RS-232, CAN, RS-422 or Ethernet, represented in Fig. 6.2.

Nowadays there exist several of-the-shelf platforms dedicated to HIL simulations, supported by powerful Digital Signal Processing architectures capable of conduct complex real-time simulations. A few examples are dSpace® [2], National Instruments® [3] and OPAL-RT® [4], currently the world leaders in this market. Besides simulating the physical processes' nominal behavior, such systems often

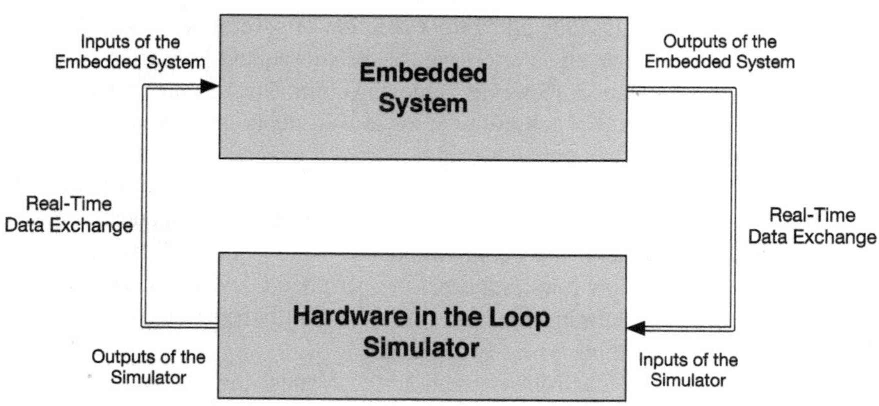

Fig. 6.1 Block diagram of embedded system connected to a HIL simulator

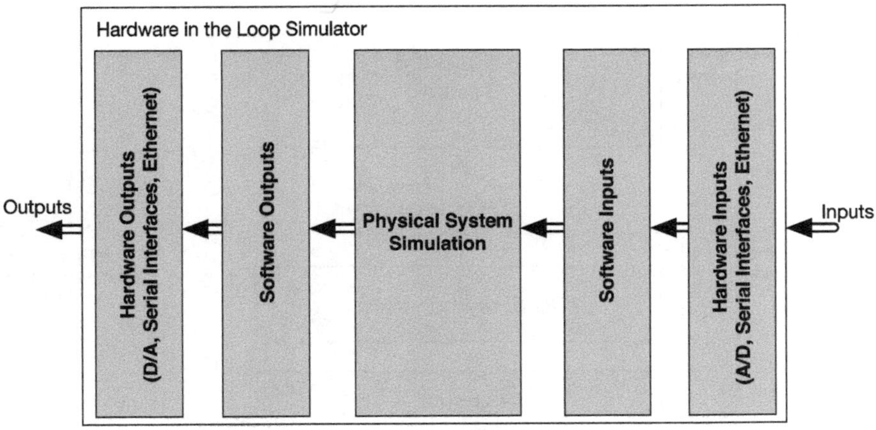

Fig. 6.2 Components of a simple HIL simulation

feature fault simulation tests which are particularly relevant for safety-critical systems in order to assess under repeatable conditions failure mode test scenarios that are difficult to conduct under real environment [2]. Complemented with simulation software such as Simulink® [5] and analog and digital I/O Front-End software as NI's LabView® [3], a HIL solution provides an efficient, reusable and safe environment where the product development can centered in the functionality of the controller without risks for either the engineer or the plant. The development of Electronic Control Units dedicated to vehicle's safety features [6] and engine control [7] in automotive industry or flight control systems in aerospace industry [8, 9] are some examples of projects whose success heavily depends on HIL's testing reliability.

Despite all the advantages, HIL's architecture also poses some limitations that should be highlighted. Firstly, since such frameworks are intended for real-time design verification, the simulated systems must consider the throughput of the HIL's processor that iterates them. For that reason, it is necessary to deterministically bound the require execution time of each simulation iteration (by using fixed-step solvers for example [2]). Consequently, highly complex process models are not adequate when a small iteration's turnaround time is required (as in high frequency control loops). Secondly, since the embedded system is decoupled from the HIL and is dependent on the simulator outputs, one cannot simple pause the simulator for in-circuit diagnosis. Hence, the full set of features provided by a HIL system are not necessarily the most adequate during earlier development stages which are more focused on the embedded system's firmware.

Processor-In-the-Loop (PIL) simulation can be considered as an intermediate stage between the traditional and HIL simulations. Similarly to the latter framework, a PIL simulation features a test environment where an embedded platform that runs the control algorithm is connected to a host computer that iterates a model of a physical process. Thus, an evaluation regarding the execution conditions of the developed

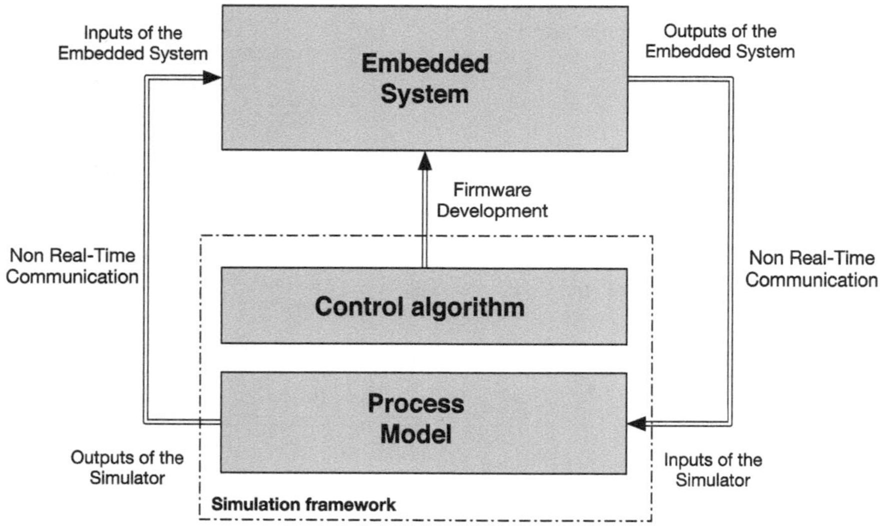

Fig. 6.3 Block diagram of embedded system connected to a PIL simulator

algorithm in a computationally constrained system can be performed, enhancing the optimization procedures for important factors such as code size, memory footprint and algorithm execution turnaround time. However, in the PIL case the simulation process is not executed in real-time, but its the pace established by the code execution time and message exchange delays between both platforms. By doing so, the validation of the developed firmware can be performed step by step comparing with the results obtained in the earlier computational simulations. The principles of PIL based development are depicted in Fig. 6.3.

Price-wise speaking, the costs of development based on a PIL framework are significantly smaller compared to the off-the-shelf HIL systems. Currently, some simulation software as Simulink® [5] and PLECS® [10] already support several development boards based on microcontrollers from the major manufacturers in the market (STM and Microchip for example), providing code generation tools [11] that significantly ease the task of converting the developed algorithms for different execution platforms.

Yet, the use of such tools present some restrictions and drawbacks, because of:

- Support only a limited set of commercially available embedded systems and development boards that may not be a perfect match for a final product
- Code generation tools most certainly will not produce the most efficient firmware for an embedded system due to the complexity of some algebras used in control algorithms
- Are mostly based on closed-source code

For those reasons and to ease the evaluation of the control algorithms implemented previously presented in an embedded platform, a PIL architecture based on the MATLAB®/Simulink® environment (for model simulation) and a embedded control system supported by the FreeRTOS® real-time kernel [12] is developed in the sequel.

6.2 PIL Architecture

According to the principles of a Processor-In-the-Loop testing framework previously presented, three main elements of its architectures must be specified, namely:

- The simulation software used by a host for the iteration of a continuous time models
- The embedded system software architecture
- The communication interface

The Simulink® toolbox [5], as part of the MATLAB® software provides an important framework for simulation of continuous time systems described by their transfer functions. Furthermore, its simulation capabilities can be significantly enhanced with the integration of additional toolboxes that already include models of complex processes and elements. They can be interconnected in a block-based approach, spanning categories such as mechanical parts, hydraulics, thermodynamic features, or electronics components, reducing the necessity of having a deep knowledge about the otherwise required mathematical models. Since Simulink® allows several methods of simulation (and among them a script based one), it is possible to implement discrete time control systems that take advantages of Simulink® continuous-time simulation—the control systems implemented in the previous chapter were simulated according to this approach. Consequently, the control algorithm can be decoupled from the model execution, easing the extension of the simulation software to include an embedded system in the loop for the implementation of the control algorithm. Bearing these ideas in mind, in the MATLAB® environment the code of control algorithm is exchanged by a communication link that transmits the plant data to the Micro-Controller Unit (MCU) and receives from it the output of the control algorithm. In the MCU side, the firmware is developed around two main tasks: receive and transmit the data over the communication link and implementation of the control algorithm. These main elements are sequentially interconnected as presented in Fig. 6.4.

6.2.1 Development Board

The implementation of versatile digital feedback control loops based on embedded systems improved the quality of many industrial processes. Still, the best performing control algorithms require computationally intensive procedures which, when executed in low-cost embedded systems, severely restrict their usability to applications

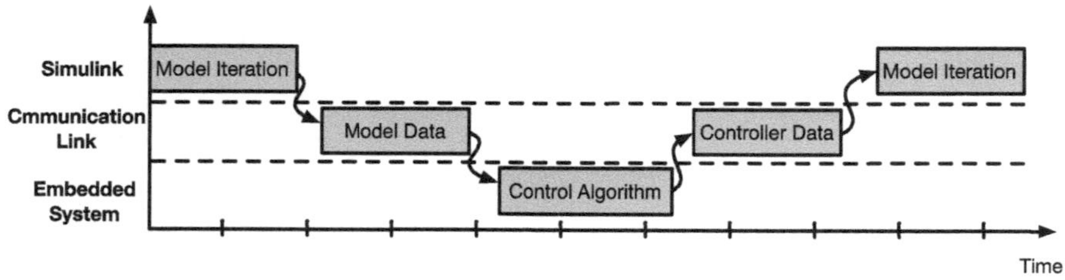

Fig. 6.4 Sequence of events during one loop of the PIL simulation

with slow dynamics to cope with the control loop calculations' turnaround time. Tackling the high performance algorithmic needs in low-cost embedded designs, ARM® recently introduced in the market a System-on-a-Chip based on a new processor family - the ARM® Cortex® M4. Complementing the available portfolio composed by the broadly used, general purpose, low-cost and low-power ARM® Cortex®-M3 and M0 families, systems based on the Cortex® M4 core stand out in the state-of-the-art of embedded systems. They feature an instruction set optimized for Digital Signal Processing operations, a single cycle Multiply and Accumulate (MAC) unit and a single-precision hardware Floating-Point Unit (FPU). The availability of a hardware FPU in such a small semiconductor die at relative low cost per unit is perhaps one of the most important enhancements of the referred architecture as it significantly simplifies the development of computationally heavy algorithms that otherwise would have to be developed through fixed-point representations. Depending on the complexity of the calculations employed in the algorithms, such procedure can become an elaborate task, requiring a deep analysis of every intermediate assessment to ensure overflow and underflow conditions will not be met during normal execution.

In order to ease the prototyping of embedded control systems, it is important to have a reusable core system featuring a basic set of peripherals devices frequently used in such applications. In the past, MCUs were often encapsulated in easy to use Dual In-line Packaging (DIP) MCUs but, currently the most powerful embedded systems are only available in high density pin-out packages which significantly increase the complexity of the earlier prototyping stages. Hence, to overcome such constraint, during the present work a system-on-a-module based on a ® Cortex® M4 MCU was designed following a DIP layout that can be easily integrated in prototype electronic systems. Figure 6.5 depicts the developed module.

The computing power of the ARM® Cortex® M4 core covers the needs of several different applications in a broad range of domains, spanning from embedded control loops to multimedia applications. For the purpose of control system's development, the NXP LPC4337 MCU is used [13] featuring a dual core architecture with a 204 MHz ARM® Cortex® M4 and a low power ARM® Cortex® M0 co-processor (which can be used to handle less demanding tasks as communications with other devices and free up the main core for real-time processing), 1 MB of flash and 136 kB on-chip SRAM, along with several configurable peripherals as two High-speed USB controllers, Ethernet, Hardware controlled Pulse Width Modulated

Fig. 6.5 Development board for embedded control based on the ARM® Cortex® M4 core

output ports, multiple communication buses and typical digital/analog ports. Hence, considering control and digital signal processing applications, the development board was designed to provide easy access to the following features:

- 16 hardware controlled PWM outputs using the Motor Control peripheral and State Configurable Timer (SCT) outputs
- 8 channels for two 10-bit ADCs and one 10-bit DAC with data conversion rate 400 kSamples/s
- Network communications based on 10/100 Ethernet link for high throughput data communications
- Two Controller Area Network (CAN), one SPI (Serial Peripheral Interface), I2C (Inter-Integrated Circuit) and one interfaces for connecting additional devices
- Integrated USB/UART converter
- High Speed USB controller with Host and Device capabilities
- One I2S (Inter-IC Sound) for connectivity with digital audio systems

To ease the interconnectivity of the board with the remaining systems, its pin-out was organized in functional groups as depicted in Fig. 6.6. It is important to note that the output ports of the board are not exclusive to the highlighted features and that it is possible to remap some of them with several other functionalities (peripherals or general purpose I/O).

6.2.2 Embedded System's Software Architecture

The increasingly computational power, memory resources and high peripheral integration available in most recent embedded systems led to significant changes in the software architecture paradigm of such small devices. In the past, embedded systems were mostly developed focusing on a particular task, but nowadays it is not uncommon to integrate several control loops and additional functionalities such as network

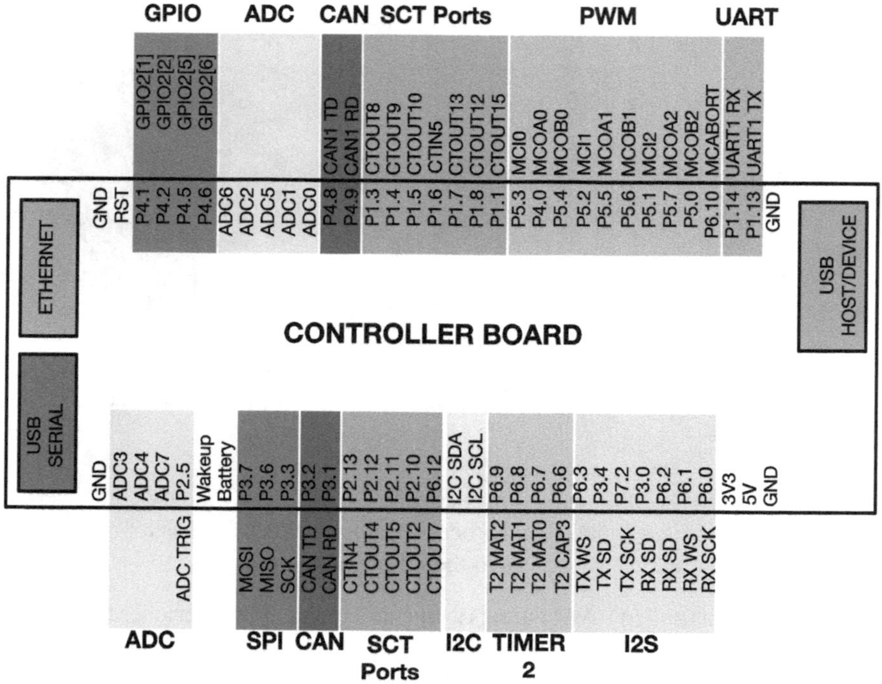

Fig. 6.6 Peripherals available in the controller board

communications in the same system which compete for processor time for their execution. In such highly integrated products, implementing software under a monolithic approach can easily become intractable for a developer. For that reason, following the longstanding practices of software development supported by an operative system, the development of an embedded system's firmware under a multi-task model with several abstraction layers has become crucial for the implementation of more complex projects. Complying with these requirements, the open-source FreeRTOS® real-time kernel [12] provides an platform agnostic Application Programming Interface (API) that establishes an abstraction layer for creating multi-task systems and managing their timings, execution priorities and inter-task communication requirements. The implemented tasks are scheduled by a tick-based fixed priority scheduler that supports preemption, which is particularly relevant for an easy development of time-triggered systems such as control loops.

Even though a program can be segmented in several tasks, in most cases there exist interdependencies among them such as execution precedences or concurrent access of shared resources as memory or peripherals. For that reason, FreeRTOS® also provides on its API synchronization mechanisms such as semaphores, mutexes and message queues to avoid race conditions between tasks that would ultimately result in an inconsistent execution of the program. Since its API is written mostly in C and is open source, this kernel is highly portable and scalable. Hence, a large portion of a system's firmware easily ported to a diverse range of embedded platforms currently supported.

6.2.2.1 PIL Integration

Using the task-based organization provided by the FreeRTOS® kernel, the embedded system's firmware can be segmented in three main objectives to be successfully integrated in the PIL framework: communication with the simulation host, update of the control system's variables and execution for the control algorithm. Following this approach, the embedded system tested in the PIL framework will require limited changes to be interfaced with a real controlled plant.

As part of a PIL system, the three implemented tasks follow a producer/consumer paradigm, existing a precedence order for their execution. Since they cannot be executed concurrently, they are created with the same priority level. Figure 6.7 depicts the sequence of events that occur at each control algorithm iteration.

Taking advantage of the high transmission rates of the Ethernet interface available in the development board, the communications between the Simulator and the Embedded System are performed over an UDP socket. Such simple transmission model encompasses a minimum set of protocol mechanisms which avoid significant overhead at network level related with message delivery failures, which is particularly suited to time-sensitive applications. In such cases, as in real-time systems, dropping the delayed packets is preferable than waiting for them. Establishing a parallelism with a deployed control system, the UDP communication task assumes the role of the ADC and DAC systems, providing the communication endpoints with the controlled process (remotely executed in this case). The System State Update task is responsible for the management of the relevant data for the controlled algorithm, as updating its regression variables (previous plant's samples and controller's outputs). Finally, the controller task implements the evaluated control algorithm. The low priority task can be used for debug purposes, as to signal stack overflow conditions or any other relevant events during the firmware development.

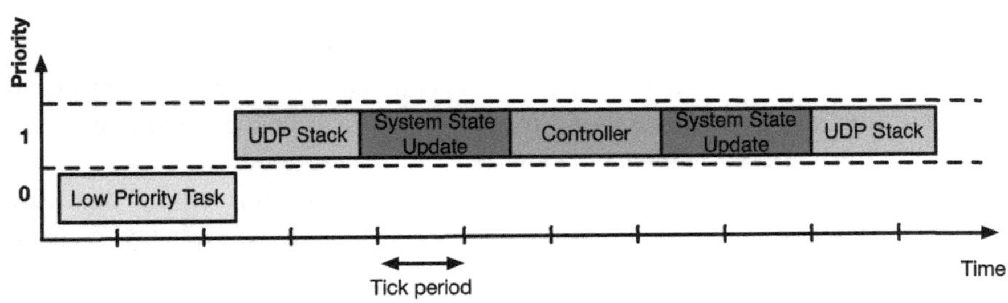

Fig. 6.7 Event triggered RTOS tasks during the Processor-In-the-Loop simulation

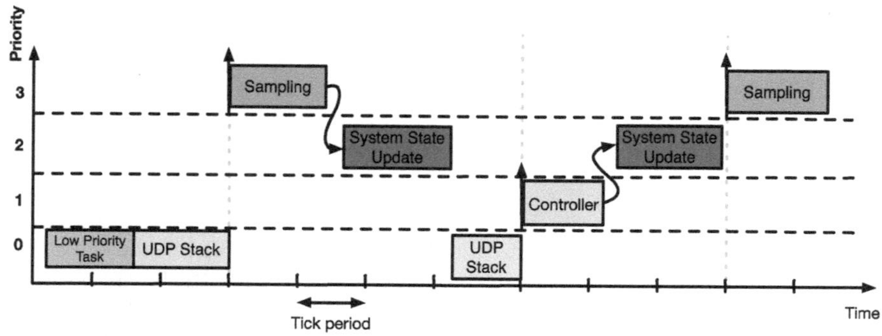

Fig. 6.8 Time triggered RTOS tasks during the real process control

Since the purpose of the PIL architecture is to provide an easy test framework for the validation of the final control system, the previously developed system can be easily extended to a time-triggered operation mode. In this scenario, the importance of the UDP stack execution is deprecated to a low priority level since its role is exchanged by a task responsible for managing the ADC readouts and control the process's actuators. Communications with a host computer can still be performed for data-logging and reconfiguration purposes which do not present any real-time requirements. As is depicted in the Fig. 6.8 the sampling task and controller tasks are now time-triggered according to the required sampling and actuation frequencies.

6.3 System Evaluation

To evaluate the PIL framework and assess the capabilities of the developed system-on-a-module, the Fermentation Reactor Temperature Control simulator previously tested will serve as benchmark for the following tests:

- Firstly, the GPC control algorithm based on the Type-2 TS FLS is executed in the embedded system and is compared with the implementation previously evaluated on simulation
- Secondly, the performance improvements obtained with the introduction of the FPU in the embedded system architecture are compared with the algorithm's turn-around time when the calculations are performed using either the hardware or software floating-point implementations
- Ultimately, the computational cost of the three GPC implementations that were used as comparison standpoint in the previous chapter are be evaluated by measuring their execution turnaround time

Similarly to the models used as support to the GPC implementations in the previous chapter, the linear approximations of the process are based on the typical second order systems's with no dead-time structure (yielding a 4 parameter linear model)

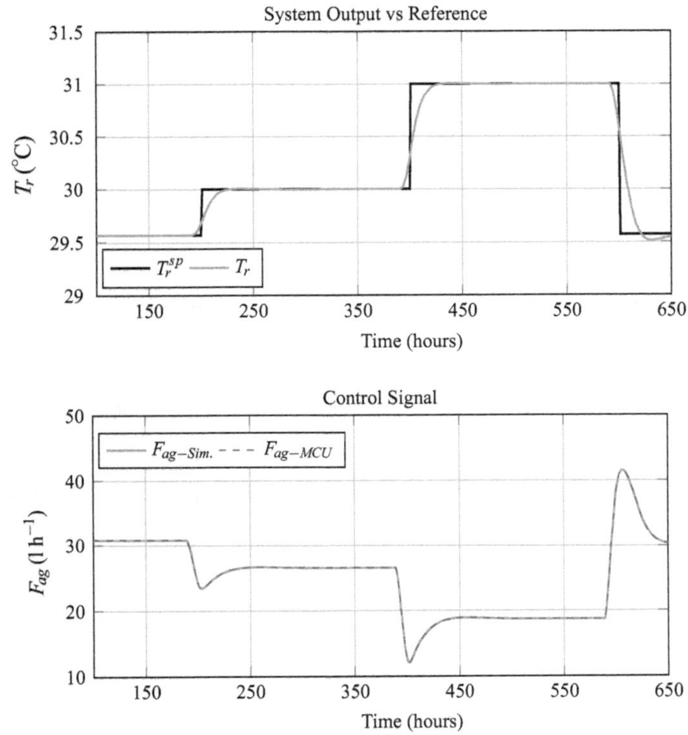

Fig. 6.9 Closed loop behavior of the reactor's temperature during a yeast fermentation reaction using a PIL simulation

and, in the cases where Type-1 and Type-2 Fuzzy Logic Models are used, a total of 5-Rules are employed to partition the model's input space.

The first evaluation scenario will highlight the PIL framework as a firmware development aiding tool. The MATLAB® work environment allows one to inspect the simulation results and to check the state of every intermediate variable, so that it is possible to verify that the values computed in the ideal simulation and in the embedded system mutually agree. In this test, the loop is closed by a GPC controller based on Type-2 TS Fuzzy Models that is executed in the MCU. As depicted in Fig. 6.9, it is seen that the control signal waveform for the MATLAB® implementation and the MCU one overlap near perfectly. Consequently, the resulting process's closed loop response is similar to the results obtained in the Chap. 5.

Although not clear at a first instance, there is a negligible difference between both controller signals as presented in Fig. 6.10. This mismatch is several orders of magnitude smaller than the signal of interest and is justifiable for accumulated errors due to the differences in the floating point representations between both architectures (MATLAB® uses the double-precision (64 bit) representation while the MCU uses single-precision one (32 bit)).

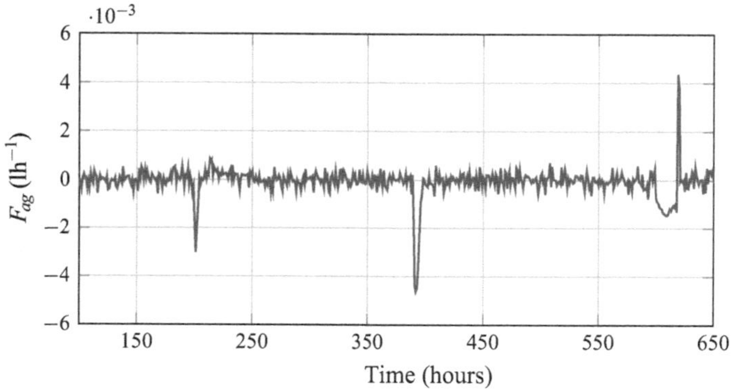

Fig. 6.10 Difference between the control signal when executed in the embedded system and in the MATLAB®

Table 6.1 Turnaround time of the predictor and control algorithm based on Type-2 Fuzzy model executed in the ARM® Cortex® M4

	Hardware FPU (μs)	Emulated FPU
GPC	59	513 μs
Type-2 Fuzzy model	110	924 μs
Total	169	1.43 ms

Putting now into perspective the importance of the hardware Cortex® M4 FPU for solving computationally intensive mathematical algorithms, the execution turnaround time for the GPC control system implementation based on a Type-2 Fuzzy Model was measured. The obtained results are presented on Table 6.1. In this test, the Cortex® M4 core was configured to run at its maximum operating frequency (204 MHz) and the measurements were taken considering the execution of the required floating-point computations, either using the standard C floating-point emulation library, or using the available hardware FPU.

Meeting our initial expectations, the introduction of the Cortex® M4 hardware FPU lead to a significant improvement of the algorithms' turnaround time, reducing it by a factor of approximately 8.5. Such achievement is important as it enables one to develop control loops for a broader range of processes. The ones with faster dynamics are particularly challenging as they demand for high frequency control loops in order correctly track their response to disturbances and operation regime's variations [14].

Comparing the computational burden of the three GPC algorithm's implementations (based on the linear ARX model, Type-1 Fuzzy model and Type-2 Fuzzy model) the turnaround time of the predictor and control algorithms was measured (when using the hardware FPU), as presented in Table 6.2.

Similarly to the results presented in Chap. 5, the GPC implementation based on the Type-2 Fuzzy Model poses a higher turnaround time due to the larger number

Table 6.2 Algorithms' turnaround time using based on different model based control implementations (N.A.—not available)

	ARX linear model	Type-1 Fuzzy model (μs)	Type-2 Fuzzy model (μs)
GPC	25.4 μs	29.2	59
Model	N/A	15.7	110
Total	25.4 μs	44.9	169

of model parameters and required calculations to obtain the control action. For the opposing reasons, the linear ARX based implementation presents the smaller computational time. Focusing firstly in the comparison between the linear ARX/ Type-1 Fuzzy Model's metrics, the major difference in the execution are due to the necessity of executing the Type-1 Fuzzy Model and performing linearization to obtain a control law. Since in the GPC implementation based on the linear ARX structure the model is considered fixed, the metrics related with the model's execution time are not available (N.A.). Considering now the two Fuzzy Model based GPC implementations, the differences in execution time are a consequence of the higher complexity of the Type-2 Fuzzy Model. The observed difference is mainly due to the Type-Reduction algorithm which takes approximately 79 μs to complete (72% of the model's execution time). The GPC algorithm takes approximately the double amount of time to execute in the Type-2 Fuzzy Model case as it effectively executed twice in order to obtain the upper and lower bounds of the control signal.

Contextualizing the measured computational time in the deployment of process's of control loops, one verifies that the use of the hardware FPU in the Cortex® M4 core significantly reduces the MCU work load, leaving a large headroom to cope with systems with faster dynamics (thus requiring a control loop with higher execution frequency) or to perform additional non real-time tasks without compromising the schedulability of the overall system. Assuming that the MCU computational power is dedicated to the control task and a DMA controller to transfer the ADC measurements to the control algorithm variables' memory region, one can develop control loops that can meet the requirements of many "fast" processes. More particularly, for the model structure considered in this evaluation, one can expect to execute a control loop at approximately 6 kHz.

6.4 Conclusions

The possibility of developing control algorithms in simulation frameworks and executing them in computationally constrained platforms such as general purpose embedded systems is of great importance for their deployment in real environments and achieve the sough performance improvements in process's manipulation. Processor-In-the Loop frameworks are undoubtedly a valuable tool supporting this transition. Yet, in many cases such step is ultimately not adopted since the complexity of the developed methods hinder their broad deployment given their superior costs of implementation. Type-2 Fuzzy Logic based systems have been several times pointed as computationally demanding but, as was assessed in this chapter, it is well under the capabilities of currently available embedded systems. The performance improvements achieved with the hardware FPU significantly contributed to the success of this analysis.

As a direct consequence of the observed results, one can say that such powerful microcontrollers will ease the development of quicker and better control loops coping with physical systems with faster dynamics. Additionally, the available performance head-room can certainly be used to deal with the overhead introduced by additional features such as a real-time kernel. Developing software in embedded systems under a monolithic approach can easily become intractable for a developer so software abstraction layers implemented by frameworks as FreeRTOS® kernel become crucial to efficiently develop multi-task systems with real-time constraints.

References

1. Schlager, M.: Hardware-in-the-Loop Simulation: A Scalable. Time-triggered Hardware-in-the-loop Simulation Framework, VDM Verlag, Component-based (2008)
2. dSpace HIL Simulation Systems, Information (2006). www.dspace.com
3. National Instruments HIL Simulation Systems, Information (2016). www.ni.com
4. OPAL-RT HIL Simulation Systems, Information (2016). http://www.opal-rt.com
5. Simulink - Simulation and Model-Based Design, Information (2016). www.mathworks.com
6. Heidrich, L., et. al.: Hardware-in-the-Loop Test Rig for Integrated Vehicle Control Systems, Advances in Automotice Control, vol. 7, part 1. In: 7th IFAC Symposium on Advances in Automotive Control (2013)
7. Poon, J., et. al.: Hardware-in-the-Loop testing for electric vehicle drive applications. In: IEEE Applied Power Electronics Conference and Exposition (APEC), pp. 2576–2582 (2012)
8. Badaruddin, K., Hernandez, J., Brown, J.: The importance of hardware-in-the-loop testing to the cassini mission to saturn. In: IEEE Aerospace Conference, pp. 1–9 (2007)
9. Guowei, G., Chen, B. M., et. al.: Design and implementation of a hardware-in-the-loop simulation system for small-scale UAV Helicopters. Mechatronics **19**(7), 1057–1066. Elsevier (2009)
10. PLECS Simulation Software for Power Electronics, Information (2006). www.plexim.com
11. Embedded-Coder MATLAB Toolbox, Information (2016). www.mathworks.com
12. FreeRTOS Cross Platform Real Time Operating System, Information (2016). www.freertos.org
13. LPC 433x Product Datasheet, Information (2016). www.nxp.com/documents/data_sheet/LPC435X_3X_2X_1X.pdf
14. Antã£o, R., Mota, A., Martins, R.: Adaptive control of a buck converter with an ARM cortex-M4. In: 16th International Power Electronics and Motion Control Conference and Exposition (PEMC), Antalya, Turkey. IEEE Conference Publications (2014)

Chapter 7
Conclusions

The knowledge embedded in Rule-Base systems, derived either from human experts or from clustering algorithms, is most of the times inconsistent due to interpersonal differences on the definition of the rule's membership functions or incomplete in some regions of the input/output space as a result from operation conditions not experienced during a model's training stage. The theory of Type-2 FLS focus on the mitigation of these problems and is already proving its advantages comparatively to alternative approaches already well established in literature.

Many improvementes of its original concepts were already proposed. Simplifications such as Interval Type-2 Fuzzy Sets and computationally efficient Type-Reduction algorithms, significantly contributed to the increase of the range of its possible application scenarios. Such changes did not compromised Type-2 FLS's main feature—embed in a compact representation the multitude of small deviations that can be defined over a single membership function.

As an incremental step over the long-standing FL theory, Type-2 FL shares many of its principles and applications. Hence, the majority of its recent publications naturally focus on the comparative analysis with its Type-1 counterpart, assessing its robustness in modeling and control applications under time-variant and noisy operation conditions. Despite all the successful implementations of model based control systems deployed both in industrial and in consumer level applications, Type-2 FL and GPC theories were up to the date two disjoint fields of expertise. Type-2 Fuzzy Control literature continues to put emphasis in traditional PID algorithms. More recently they became significantly biased towards the use of computationally intensive Genetic and other Bio-Inspired optimization. In a sense of control theory fundamentals, such approaches "blindly" seek the optimal controller parameters to achieve the best input/output behavior. While at first instance model based control algorithms present the developer with complex algebraic formulations to attain a control law, ultimately they become simpler to implement, are more predictable and provide well performing controllers which are better suited for real world applications. For that reason, this book proposed the development of a control system based on Generalized Predictive Control algorithms and Interval Type-2 Takagi-Sugeno Fuzzy Models.

R. Antão, *Type-2 Fuzzy Logic*, Nonlinear Physical Science,
DOI 10.1007/978-981-10-4633-9_7

In what regards to the model development procedures, as discussed in Chap. 3, the ability of keeping a meaningful model representation after the training stage is important not only for its interpretability but also to ensure the robustness of the subsequent controller synthesis procedures. Comparatively to Type-1 Fuzzy Models, Type-2 TS inherently present a greater number of tunable parameters which grant them superior levels of adaptation but, at the same time, require greater care during the training stage. Since a Type-2 Fuzzy Set representation inherently establishes an interdependency between its tunable parameters (its uncertainty bounds are defined based on a spread factor over a nominal value), the training mechanisms must ensure that such parameters are not driven in significantly different directions, ultimately disrupting the concept of a Footprint of Uncertainty. Unfortunately, under long training procedures such scenario is not so uncommon. Therefore, in order to overcome such limitation, the methods used in this book focused on finding the approximate centers of such FOU, introducing afterwards the uncertainty factors. As it was shown in this book, a relatively small "fuzzily" defined FOU used in combination with a Type-Reduction algorithm already introduces significant changes in the input/output relationships of the model, yielding improved results without necessarily increase the model dimensionality. Following such approach provides the user a deeper insight about the influence of the uncertainty factors on the quality and accuracy of the developed model. Furthermore, it makes the training algorithm less demanding in a computational sense given the smaller number of parameters that must be tuned. Ultimately, it improves the dynamic model's numerical robustness since a smaller number of degrees of freedom reduce the chances of the optimization problem to diverge to unwanted solutions or be trapped in a local minima.

As was debated in the Chaps. 4 and 5, the efficiency of a GPC implementation is ultimately defined by the accuracy of the model used for approximating the expected future behavior of a physical process. Since in some applications the model mismatch is significant, the control law obtained is not capable to ensure the quality metrics sought for the closed loop system. To overcome such limitation, non-linear GPC implementations received in recent year a crescent interest from researchers but still, methods based on linear approximations of the processes are of great value for industry due to the computational efficiency of the closed form algebraic methods required to solve them. Takagi-Sugeno Fuzzy Models stand as a particular type of structure that complies with both requirements and, by extending it with the Type-2 Fuzzy Logic formalisms, one aimed to improve the accuracy of its locally linearized models to extrapolate better n-step ahead predictors. Such improvements are particularly important during abrupt changes on the operation regimes, transients where the overall model results from the contribution of several locally linear sub-models with smaller validity and, consequently, the uncertainty over the obtained predictions is inherently higher. Based on the results attained, it was observed that, at the cost of a small increase of the computational effort, a Generalized Predictive Controller based on Type-2 FLSs presents an improved transient behavior comparatively to its Type-1 and linear ARX counterparts under similar operation conditions, more particularly when the system is subject to unmodeled disturbances. Although the computational time was not a limitation factor in the presented scenarios, the improvements achieved with the proposed method can be extended to non-linear systems with smaller time constants as well, proving itself as a valid alternative approach to non-linear model predictive control that require the use of heavy non-linear optimization algorithms at every control interval. The analysis performed only considered systems up to second order and without dead-time. Thus, the tests under different conditions remain for a future evaluation.

Even though the proposed control framework was developed and evaluated under simulated conditions, the data samples used during the model extraction procedures and the closed loop test scenarios were corrupted by gaussian noise, introducing disturbances with amplitudes similar to those possibly occuring in real operation conditions. Hence, together with the developed Processor-in-the-Loop framework, it is expected that the time of deployment of the proposed control system can be significantly reduced in future works inspired by this book, thus taking the state-of-the-art of Type-2 Fuzzy Logic one little step forward.

Appendix

Included Software

This book is accompanied with a software framework that provides the test scenarios evaluated in this book and, more importantly, allows complementary work to be developed based on the proposed principles. The materials comprise examples for the Matlab platform and a C implementation for embedded systems.

Regarding the Matlab implementations, no additional toolboxes are required and 4fourindependent projects are provided, namelly:

- ARX—Reactor: GPC implementation based on linear ARX model of the Stirred Reactor temperature control process.
- ARX—Tank: GPC implementation based on linear ARX model of the Coupled Tanks liquid level control process.
- TS—Reactor: GPC implementation based on Takagi Sugeno Fuzzy models (Type-1 and Type-2) of the Stirred Reactor temperature control process.
- TS—Tank: GPC implementation based on Takagi Sugeno Fuzzy models (Type-1 and Type-2) of the Coupled Tanks liquid level control process.

The referred examples include the datasets and identified models' parameters (inside the *data* folder), so the closed loop control scripts can be straightforwardly evaluated. Nevertheless, the reader can generate new datasets and perform new model identifications using the *openloop* and *getmodel* scripts. If so, to achieve proper control results, a successful model identification is required thus the identification stages must provide proper excitation of the model over its different operation ranges and controllable regimes due to their non-linear nature (as described in Chap. 4). This is required so each individual model parameters' converge to their appropriate values.

R. Antão, *Type-2 Fuzzy Logic*, Nonlinear Physical Science,
DOI 10.1007/978-981-10-4633-9

Additionally, a generic library written in C is distributed for embedded systems implementation. The example of the control task (*ctrltask.c*) exemplifies the sequence of actions to implement the closed loop controller. It is written using the FreeRTOS API but can easily be ported to any other embedded platform. The file *t2fls.h* allows the user to parametrize the model by the number of regression variables, number of rules and its antecedents, while the *gpc.h* sets the parameters related with the control algorithm as its control and estimation windows.

The companion software is available at: http://extras.springer.com

FSC
www.fsc.org
MIX
Papier | Fördert
gute Waldnutzung
FSC® C083411

Kolibri 360 GmbH
Ferdinand-Jühlke-Straße 7
99095 Erfurt, Deutschland
produktsicherheit@kolibri360.de